I0106142

MEP 804A/B AND 814A/B
15 KW GENERATOR SET
REPAIR PARTS
AND SPECIAL TOOLS LISTS
TM 9-6115-643-24P

FOR

GENERATOR SET, SKID MOUNTED, TACTICAL QUIET,
15 KW, 50/60 HZ
MEP-804A
(NSN: 6115-01-274-7388)
MEP-804B
(NSN: 6115-01-530-1458)
GENERATOR SET, SKID MOUNTED, TACTICAL QUIET,
15 KW, 400 HZ
MEP-814A
(NSN: 6115-01-274-7393)
MEP-814B
(NSN: 6115-01-529-9494)

edited by
Brian Greul

The MEP series of Military Generators are reknowned for their quiet, durable operation and conservative power ratings. This is the parts and special tools list (manual) for the 15KW version of the generator issued under models 804 and 814. The A series has analog controls and the B series has digital controls. Various units are manufactured for the US Government by different contractors with different power plants. This book is a reprint of the manual published by the US Army. It is printed as a courtesy to enthusiasts and owners of these generator sets. Other important manuals for this generator are also printed by this publisher.

An 8.5x11 3 hole punched loose leaf copy may be purchased for your 3 ring binder. Email books@ocotillopress.com for current information.

Should you have suggestions or feedback on ways to improve this book please send email to Books@OcotilloPress.com We also welcome requests for other military manuals you would like to see printed.

Edited 2021 Ocotillo Press
ISBN 978-1-954285-16-3

Printed in the United States of America

Ocotillo Press
Houston, TX 77017
Books@OcotilloPress.com

*ARMY TM 9-6115-643-24P
AIR FORCE TO 35C2-3-445-24

TECHNICAL MANUAL

REPAIR PARTS AND SPECIAL TOOLS LISTS
FOR

GENERATOR SET, SKID MOUNTED, TACTICAL QUIET, 15 kW, 50/60 Hz

MEP-804A
(NSN: 6115-01-274-7388)
MEP-804B
(NSN: 6115-01-530-1458)

GENERATOR SET, SKID MOUNTED, TACTICAL QUIET, 15 kW, 400 Hz

MEP-814A
(NSN: 6115-01-274-7393)
MEP-814B
(NSN: 6115-01-529-9494)

*SUPERSEDURE NOTICE - This manual supersedes TM 9-6115-643-24P dated 15 November 2008.

HEADQUARTERS, DEPARTMENTS OF THE ARMY AND AIRFORCE

LIST OF EFFECTIVE PAGES / WORK PACKAGES

NOTE: This manual supersedes TM 9-6115-643-24P dated 15 November 2008. Zero in the “Change No.” column indicates an original page or work package.

Date of issue for the original manual is:

Original 30 JULY 2010

TOTAL NUMBER OF PAGES FOR FRONT AND REAR MATTER IS 8 AND TOTAL NUMBER OF WORK PACKAGES IS 44, CONSISTING OF THE FOLLOWING:

Page / WP No.	Change No.	Page / WP No.	Change No.
Front Cover	0	WP 0037 (4 pgs)	0
Blank	0	WP 0038 (4 pgs)	0
i- iv	0	WP 0039 (4 pgs)	0
Chp 1 title page	0	WP 0040 (4 pgs)	0
Chp 1 Index	0	WP 0041 (2 pgs)	0
WP 0001 (6 pgs)	0	WP 0042 (2 pgs)	0
WP 0002 (4 pgs)	0	WP 0043 (14 pgs)	0
WP 0003 (9 pgs)	0	WP 0044 (20 pgs)	0
WP 0004 (4 pgs)	0	Inside back cover	0
WP 0005 (4 pgs)	0	Back cover	0
WP 0006 (4 pgs)	0		
WP 0007 (4 pgs)	0		
WP 0008 (6 pgs)	0		
WP 0009 (36 pgs)	0		
WP 0010 (4 pgs)	0		
WP 0011 (4 pgs)	0		
WP 0012 (4 pgs)	0		
WP 0013 (4 pgs)	0		
WP 0014 (6 pgs)	0		
WP 0015 (4 pgs)	0		
WP 0016 (6 pgs)	0		
WP 0017 (4 pgs)	0		
WP 0018 (4 pgs)	0		
WP 0019 (12 pgs)	0		
WP 0020 (6 pgs)	0		
WP 0021 (4 pgs)	0		
WP 0022 (10 pgs)	0		
WP 0023 (4 pgs)	0		
WP 0024 (4 pgs)	0		
WP 0025 (4 pgs)	0		
WP 0026 (6 pgs)	0		
WP 0027 (6 pgs)	0		
WP 0028 (4 pgs)	0		
WP 0029 (4 pgs)	0		
WP 0030 (4 pgs)	0		
WP 0031 (4 pgs)	0		
WP 0032 (4 pgs)	0		
WP 0033 (4 pgs)	0		
WP 0034 (4 pgs)	0		
WP 0035 (4 pgs)	0		
WP 0036 (4 pgs)	0		

HEADQUARTERS,
DEPARTMENTS OF THE ARMY AND AIR FORCE
WASHINGTON, D.C.,30 JULY 2010

TECHNICAL MANUAL

REPAIR PARTS AND SPECIAL TOOLS LISTS FOR

GENERATOR SET, SKID MOUNTED, TACTICAL QUIET, 15 kW, 50/60 Hz

MEP-804A
(NSN: 6115-01-274-7388)
MEP-804B
(NSN: 6115-01-530-1458)

GENERATOR SET, SKID MOUNTED, TACTICAL QUIET, 15 kW, 400 Hz

MEP-814A
(NSN: 6115-01-274-7393)
MEP-814B
(NSN: 6115-01-529-9494)

REPORTING ERRORS AND RECOMMENDING IMPROVEMENTS

You can help improve this manual. If you find any mistakes or if you know of a way to improve the procedures, please let us know. Reports, as applicable by the requiring Service, should be submitted as follows:

(a) (A) Army - Mail your letter or DA Form 2028 (Recommended Changes to Publications and Blank Forms), located in the back of this manual, directly to: Commander, U.S. Army CECOM (LCMC) and Fort Monmouth, ATTN: AMSEL-LC-LEO-E-CM, Fort Monmouth, NJ 07703-5006. You may also send in your recommended changes via electronic mail or by fax. Our fax number is 732-532-3421, DSN 992-3421. Our e-mail address is MONM-AMSELLEOPUBSCHG@conus.army.mil. Our online web address for entering and submitting DA Form 2028s is http://edm.monmouth.army.mil/pubs/2028.html.

(b) (F) Air Force - By Air Force AFTO Form 22 (Technical Manual (TM) change Recommendation and Reply) in accordance with paragraph 6-5, Section VI, TO 00-5-1 directly to prime ALC/MST. A reply will be furnished to you.

***SUPERSEDURE NOTICE** - This manual supersedes TM 9-6115-643-24P dated 15 November 2008.

TABLE OF CONTENTS

Page No.
WP Sequence No.

TABLE OF CONTENTS - Continued

Page No.
WP Sequence No.

TABLE OF CONTENTS - Continued

Page No.
WP Sequence No.

CHAPTER 1

FIELD AND SUSTAINMENT REPAIR PARTS AND SPECIAL TOOLS LIST

FOR

GENERATOR SET, SKID MOUNTED, TACTICAL QUIET
15 kW, 50/60 Hz
MEP-804A
MEP-804B

15 kW, 400 Hz
MEP-814A
MEP-814B

CHAPTER 1

PARTS INFORMATION

WORK PACKAGE INDEX

Title WP Sequence No.

Title WP Sequence No.

FIELD AND SUSTAINMENT MAINTENANCE

15 kW 50/60 AND 400 Hz SKID MOUNTED TACTICAL QUIET GENERATOR SETS

INTRODUCTION

INTRODUCTION

SCOPE

This RPSTL lists and authorizes spares and repair parts; special tools; special test, measurement, and diagnostic equipment (TMDE); and other special support equipment required for performance of field and sustainment maintenance of the generator set. It authorizes the requisitioning, issue, and disposition of spares, repair parts, and special tools as indicated by the source, maintenance, and recoverability (SMR) codes.

GENERAL

In addition to the Introduction work package, this RPSTL is divided into the following work packages.

1. Repair Parts List Work Packages. Work packages containing lists of spares and repair parts authorized by this RPSTL for use in the performance of maintenance. These work packages also include parts which must be removed for replacement of the authorized parts. Parts lists are composed of functional groups in ascending alphanumeric sequence, with the parts in each group listed in ascending figure and item number sequence. Sending units, brackets, filters, and bolts are listed with the component they mount on. Bulk materials are listed by item name in FIG. BULK at the end of the work packages. Repair parts kits are listed separately in their own functional group and work package Repair parts for reparable special tools are also listed in a separate work package. Items listed are shown on the associated illustrations.
2. Special Tools List Work Packages. Work packages containing lists of special tools, special TMDE, and special support equipment authorized by this RPSTL (as indicated by Basis of Issue (BOI) information in the DESCRIPTION AND USABLE ON CODE (UOC) column). Tools that are components of common tool sets and/ or Class VII are not listed.
3. Cross-Reference Indexes Work Packages. There are 3 crossreference indexes work packages in this RPSTL: the National Stock Number (NSN) Index work package, the Part Number (P/N) Index work package and a Figure and Item Number Index work package. The National Stock Number Index work package refers you to the figure and item number. The Part Number Index work package refers you to the figure and item number.

EXPLANATION OF COLUMNS IN THE REPAIR PARTS LIST AND SPECIAL TOOLS LIST WORK PACKAGES

ITEM NO. (Column (1)). Indicates the number used to identify items called out in the illustration.

SMR CODE (Column (2)). The SMR code containing supply/requisitioning information, maintenance level authorization criteria, and disposition instruction, as shown in the following breakout. This entry may be subdivided into 4 subentries, one for each service.

Table 1. SMR Code Explanation.

Source Code Maintenance Code Recoverability Code

XX XX X 1st two positions: 3rd position: 4th position: 5th position:

How to get an item. Who can install, replace, or use the item. Who can do complete repair* on the item. Who determines dispo-si-tion action on unser-

*Complete Repair: Maintenance capacity, capability, and authority to perform all corrective maintenance tasks of the "Repair" function in a use/user environment in order to restore serviceability to a failed item.

Source Code. The source code tells you how you get an item needed for maintenance, repair, or overhaul of an end item/equipment. Explanations of source codes follow:

Source Code	Application/Explanation
PA PB PC PD PE PF PG PH PR PZ	**NOTE** Items coded PC are subject to deterioration. Stock items; use the applicable NSN to requisition/ request items with these source codes. They are authorized to the level indicated by the code entered in the third position of the SMR code.
KD KF KB	Items with these codes are not to be requested/ requisitioned individually. They are part of a kit which is authorized to the maintenance level indicated in the third position of the SMR code. The complete kit must be requisitioned and applied.
MF-Made at field MH-Made at below depot/sustainment level ML-Made at SRA MD-Made at depot MG-Navy only	Items with these codes are not to be requisitioned/ requested individually. They must be made from bulk material which is identified by the part number in the DESCRIPTION AND USABLE ON CODE (UOC) column and listed in the bulk material group work package of the RPSTL. If the item is authorized to you by the third position code of the SMR code, but the source code indicates it is made at higher level, order the item from the higher level of maintenance.
AF-Assembled by field AH-Assembled by below depot sustainment level AL-Assembled by SRA AD-Assembled by depot AG-Navy only	Items with these codes are not to be requested/ requisitioned individually. The parts that make up the assembled item must be requisitioned or fabricated and assembled at the level of maintenance indicated by the source code. If the third position of the SMR code authorizes you to replace the item, but the source code indicates the item is assembled at a higher level, order the item from the higher level of maintenance.
XA	Do not requisition an “XA” coded item. Order the next higher assembly. (Refer to NOTE below.)
XB	If an item is not available from salvage, order it using the CAGEC and part number.
XC	Installation drawings, diagrams, instruction sheets, field service drawings; identified by manufacturer’s part number.
XD	Item is not stocked. Order an XD-coded item through local purchase or normal supply channels using the CAGEC and part number given, if no NSN is available.

NOTE

Cannibalization or controlled exchange, when authorized, may be used as a source of supply for items with the above source codes except for those items source coded "XA" or those aircraft support items restricted by requirements of AR 750-1.

Maintenance Code. Maintenance codes tell you the level(s) of maintenance authorized to use and repair support items. The maintenance codes are entered in the third and fourth positions of the SMR code as follows:

Third Position. The maintenance code entered in the third position tells you the lowest maintenance level authorized to remove, replace, and use an item. The maintenance code entered in the third position will indicate authorization to the following levels of maintenance:

Maintenance Code Application/Explanation

F - Field maintenance can remove, replace, and use the item.

H - Below Depot Sustainment maintenance can remove, replace, and use the item.

L - Specialized repair activity can remove, replace, and use the item.

G - Afloat and ashore intermediate maintenance can remove, replace, and use the item (Navy only)

K - Contractor facility can remove, replace, and use the item.

Z - Item is not authorized to be removed, replaced, or used at any maintenance level D - Depot can remove, replace, and use the item.

*NOTE - Army may use C in the third position. However, for joint service publications, Army will use O.

Fourth Position. The maintenance code entered in the fourth position tells you whether or not the item is to be repaired and identifies the lowest maintenance level with the capability to do complete repair (perform all authorized repair functions).

NOTE

Some limited repair may be done on the item at a lower level of maintenance, if authorized by the Maintenance Allocation Chart (MAC) and SMR codes.

Maintenance Code Application/Explanation

F - Field is the lowest level that can do complete repair of the item.

H - Below Depot Sustainment is the lowest level that can do complete repair of the item.

L - Specialized repair activity is the lowest level that can do complete repair of the item.

D - Depot is the lowest level that can do complete repair of the item.

G - Both afloat and ashore intermediate levels are capable of complete repair of item. (Navy only)

K - Complete repair is done at contractor facility

Z - Nonreparable. No repair is authorized.

B - No repair is authorized. No parts or special tools are authorized for maintenance of "B" coded item. However, the item may be reconditioned by adjusting, lubricating, etc., at the user level.

Recoverability Code. Recoverability codes are assigned to items to indicate the disposition action on unserviceable items. The recoverability code is shown in the fifth position of the SMR code as follows:

Recoverability Code	Application/Explanation
Z - Nonreparable item.	When unserviceable, condemn and dispose of the item at the level of maintenance shown in the third position of the SMR code.
F - Reparable item.	When uneconomically reparable, condemn and dispose of the item at the field level.
H - Reparable item.	When uneconomically reparable, condemn and dispose of the item at the below depot sustainment level.
D - Reparable item.	When beyond lower level repair capability, return to depot. Condemnation and disposal of item are not authorized below depot level.
L - Reparable item.	Condemnation and disposal not authorized below Specialized Repair Activity (SRA).
A - Item requires special handling or condemnation procedures because of specific reasons	(such as precious metal content, high dollar value, critical material, or hazardous material). Refer to appropriate manuals/directives for specific instructions.
G - Field level reparable item.	Condemn and dispose at either afloat or ashore intermediate levels. (Navy only)
K - Reparable item.	Condemnation and disposal to be performed at contractor facility.

NSN (Column (3)). The NSN for the item is listed in this column.

CAGEC (Column (4)). The Commercial and Government Entity Code (CAGEC) is a five-digit code which is used to identify the manufacturer, distributor, or Government agency/activity that supplies the item.

PART NUMBER (Column (5)). Indicates the primary number used by the manufacturer (individual, company, firm, corporation or Government activity), which controls the design and characteristics of the item by means of its engineering drawings, specifications, standards, and inspection requirements to identify an item or range of items.

NOTE

When you use an NSN to requisition an item, the item you receive may have a different part number from the number listed.

DESCRIPTION AND USABLE ON CODE (UOC) (Column (6)). This column includes the following information:

1. The federal item name, and when required, a minimum description to identify the item.
2. Part numbers of bulk materials are referenced in this column in the line entry to be manufactured or fabricated.
3. Hardness Critical Item (HCI). A support item that provides the equipment with special protection from electromagnetic pulse (EMP) damage during a nuclear attack.
4. The statement END OF FIGURE appears just below the last item description in column (6) for a given figure in both the repair parts list and special tools list work packages.

QTY (Column (7)). The QTY (quantity per figure) column indicates the quantity of the item used in the breakout shown on the illustration/figure, which is prepared for a functional group, subfunctional group, or an assembly. A "V" appearing in this column instead of a quantity indicates that the quantity is variable and quantity may change from application to application.

EXPLANATION OF CROSS-REFERENCE INDEXES WORK PACKAGES FORMAT AND COLUMNS

1. National Stock Number (NSN) Index Work Package. NSN's in this index are listed in National Item Identification Number (NIIN) sequence.

   STOCK NUMBER Column. This column lists the NSN in NIIN sequence. The NIIN consists of the last nine digits of the NSN. When using this column to locate an item, ignore the first four digits of the NSN. However, the complete NSN should be used when ordering items by stock number.

   For example, if the NSN is 5385-01-574-1476, the NIIN is 01-574-1476.

FIG. Column. This column lists the number of the figure where the item is identified/located. The figures are in numerical order in the repair parts list and special tools list work packages.

ITEM Column. The item number identifies the item associated with the figure listed in the adjacent FIG. column. This item is also identified by the NSN listed on the same line.

2. Part Number (P/N) Index Work Package. Part numbers in this index are listed in ascending alphanumeric sequence (vertical arrangement of letter and number combinations which places the first letter or digit of each group in order A through Z, followed by the numbers 0 through 9 and each following letter or digit in like order).

PART NUMBER Column. Indicates the part number assigned to the item.

FIG. Column. This column lists the number of the figure where the item is identified/located in the repair parts list and special tools list work packages.

ITEM Column. The item number is the number assigned to the item as it appears in the figure referenced in the adjacent figure number column.

SPECIAL INFORMATION

UOC. The UOC appears in the lower left corner of the Description Column heading. Usable on codes are shown as "UOC: ..." in the Description Column (justified left) on the first line under the applicable item/nomenclature. Uncoded items are applicable to all models. Identification of the UOCs used in the RPSTL are:

Code Used On

EMK 15 kW, 50/60 Hz, TQG Set, Model MEP-804A 86Q 15 kW,

50/60 Hz, TQG Set, Model MEP-804B YNN 15 kW, 400 Hz, TQG

Set, Model MEP-814A 86R 15 kW, 400 Hz, TQG Set, Model MEP-

814B

Fabrication Instructions. Bulk materials required to manufacture items are listed in the bulk material functional group of this RPSTL. Part numbers for bulk material are also referenced in the Description Column of the line item entry for the item to be manufactured/fabricated.

Index Numbers. Items which have the word BULK in the figure column will have an index number shown in the item number column. This index number is a cross-reference between the NSN / Part Number (P/N) Index work packages and the bulk material list in the repair parts list work package.

Associated Publications. The publication(s) listed below pertains to the generator set:

Publication Short Title

LO 9-6115-643-12 Lubrication Order, 15 kW TQG Set

TM 9-6115-643-10 Operator's Manual, 15 kW TQG Set

TM 9-6115-643-24 Maintenance Manual, 15 kW TQG Set

TM 9-2815-254-24 Maintenance Manual, Diesel Engine, Model No. C-240PW-28

TM 9-2815-254-24P RPSTL, Diesel Engine, Model No. C-240PW-28

TM 9-2815-538-24&P Maintenance Manual with RPSTL, Diesel Engine, Model No. 4TNV84T-DFM

Illustrations List. The illustrations in this RPSTL contain field authorized items. The tabular list in the repair parts list work package contains only those parts coded "F" in the third position of the SMR code, therefore, there may be a break in the item number sequence.

HOW TO LOCATE REPAIR PARTS

1. When NSNs or Part Numbers Are Not Known.

    First. Using the table of contents, determine the assembly group to which the item belongs. This is necessary since figures are prepared for assembly groups and subassembly groups, and lists are divided into the same groups.

    Second. Find the figure covering the functional group or the subfunctional group to which the item belongs.

    Third. Identify the item on the figure and note the number(s).

    Fourth. Look in the repair parts list work packages for the figure and item numbers. The NSNs and part numbers are on the same line as the associated item numbers.

2. When NSN Is Known.

    First. If you have the NSN, look in the STOCK NUMBER column of the NSN index work package. The NSN is arranged in NIIN sequence. Note the figure and item number next to the NSN.

    Second. Turn to the figure and locate the item number. Verify that the item is the one you are looking for.

3. When Part Number Is Known.

    First. If you have the part number and not the NSN, look in the PART NUMBER column of the part number index work package. Identify the figure and item number.

    Second. Look up the item on the figure in the applicable repair parts list work package.

ABBREVIATIONS

Abbreviation	Explanation

BOI Basis Of Issue

CAGEC Commercial And Government Entity Code

DS Direct Support

GS General Support

MAC Maintenance Allocation Chart

NIIN National Item Identification Number (consists of the last 9-digits of the NSN) NSN National Stock Number

RPSTL Repair Parts and Special Tools List

SMR Source, Maintenance and Recoverability Code

SRA Specialized Repair Activity

TMDE Test, Measurement and Diagnostic Equipment

UOC Usable On Code

END OF WORK PACKAGE

FIELD AND SUSTAINMENT MAINTENANCE

15 kW 50/60 AND 400 Hz SKID MOUNTED TACTICAL QUIET GENERATOR SETS

GROUP 01 DC ELECTRICAL SYSTEM: BATTERIES AND CABLES

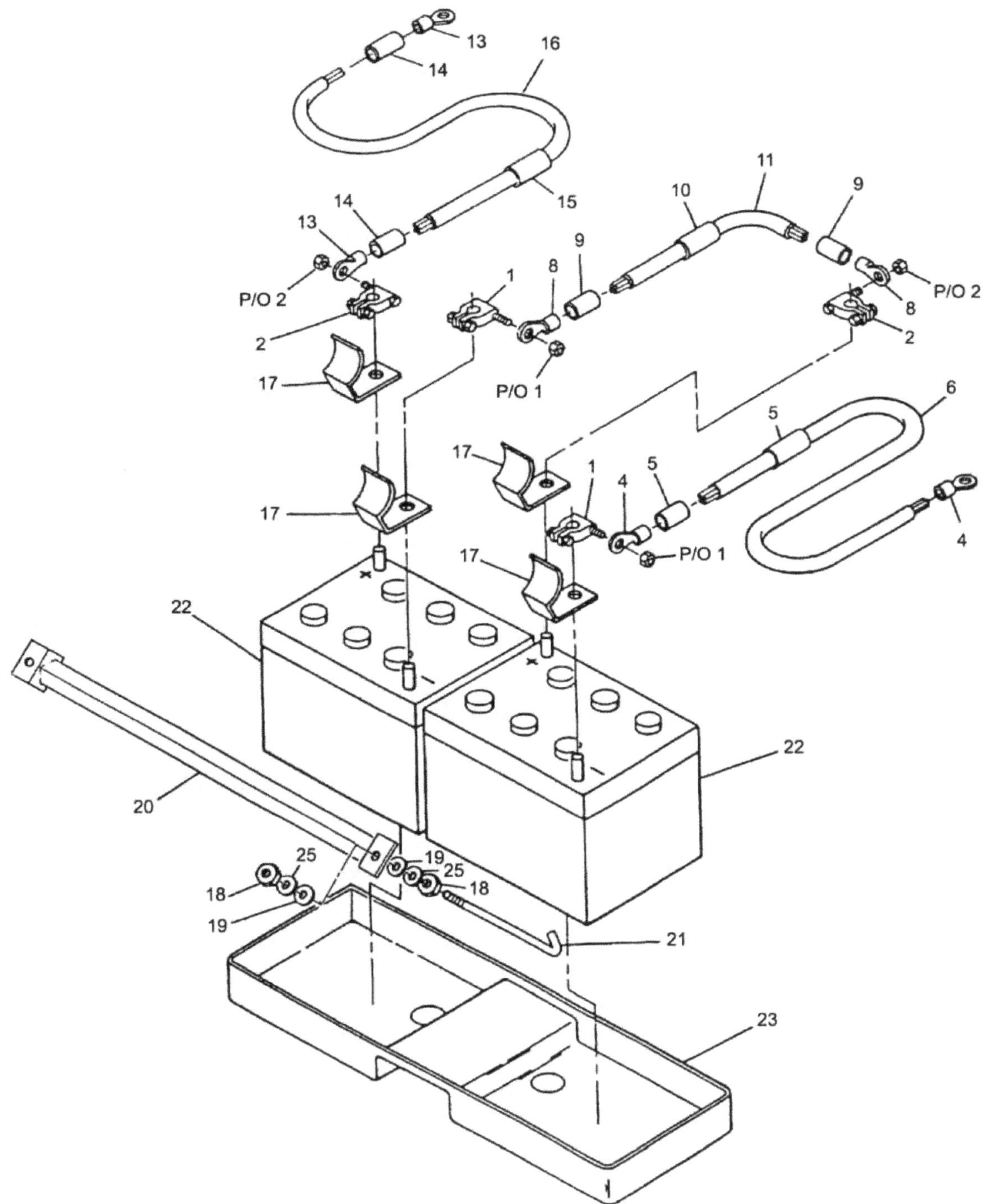

Figure 1. Batteries and Cables (Sheet 1 of 2).

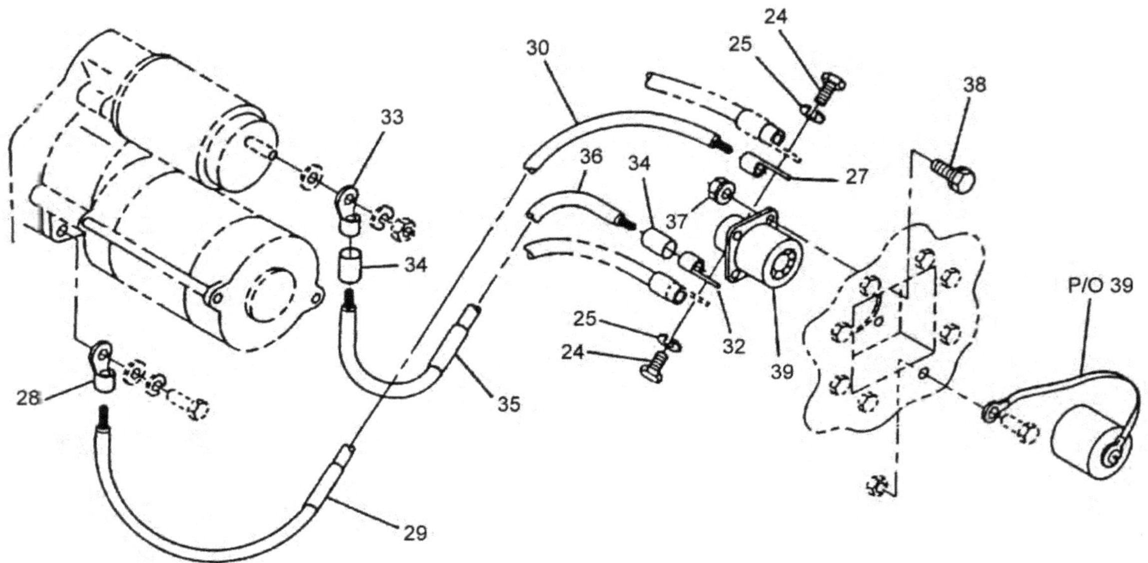

Figure 1. Batteries and Cables (Sheet 2 of 2).

(1)	(2)		(3)	(4)	(5)	(6)	(7)
	SMR CODE						
	a.	b.					
ITEM NO	ARMY	AIR FORCE	NSN	CAGEC	PART NUMBER	DESCRIPTION AND USABLE ON CODE (UOC)	QTY

GROUP 01 DC ELECTRICAL
SYSTEM

FIG. 1 BATTERIES AND CABLES

1 PAFZZ PAOZA 5940-00-549-6583 58536 AA-52425-2 . ADAPTER, BATTERY TER NEGAT-
IVE 2
2 PAFZZ PAOZZ 5940-00-549-6581 58536 A52425-1 . ADAPTER, BATTERY TER
POSITIVE 2
3 AFFFF AOO 6150-00-549-6581 30554 88-22123 . CABLE, BATTERY (NOT SHOWN) 1
4 PAFZZ PAOZZ 5940-01-369-2872 30554 88-22119-14 . . TERMINAL, LUG .... 2 5 MFFZZ MOO 30554 88-22123-3
. . INSULATION, SLEEVING MAKE
FROM P/N M23053/5-110-9 (81349),
AS REQUIRED 2
6 MFFZZ MOO 30554 88-22123-1 . . WIRE, ELECTRIC MAKE FROM P/N
J1127SGR1AWG BLK (81343), 18.0
INCHES REQUIRED 1
7 AFFFF AOO 6140-01-418-6342 30554 88-22179 . LEAD, STORAGE, BATTERY (NOT
SHOWN) 1
8 PAFZZ PAOZZ 5940-01-369-2872 30554 88-22119-14 . . TERMINAL, LUG .... 2 9 MFFZZ MOO 30554 88-22179-3
. . INSULATION, SLEEVING MAKE
FROM P/N M23053/5-110-2 (81349),
AS REQUIRED 2
10 MFFZZ MOO 30554 88-22179-4 . . INSULATION, SLEEVING MAKE
FROM P/N M23053/5-110-9 (81349),
AS REQUIRED 1
11 MFFZZ MOO 30554 88-22179-1 . . WIRE, ELECTRIC MAKE FROM P/N
J1127SGR1AWG BLK (81343), 14.0
INCHES REQUIRED 1
12 AFFFF AOO 6150-01-494-3983 30554 88-22310 . BATTERY, CABLE (NOT SHOWN) 1
13 PAFZZ PAOZZ 5940-01-369-2872 30554 88-22119-14 . . TERMINAL, LUG .... 2 14 MFFZZ MOO 30554 88-
22310-3 . . INSULATION, SLEEVING MAKE
FROM P/N M23053/5-110-2 (81349),
AS REQUIRED 2
15 MFFZZ MOO 30554 88-22310-4 . . INSULATION, SLEEVING MAKE
FROM P/N M23053/5-10-9 (81349), AS
REQUIRED 1
16 MFFZZ MOO 30554 88-22310-1 . . WIRE, ELECTRIC MAKE FROM P/N
J1127SGR1AWGBLK (81343), 29.0
INCHES REQUIRED 1
17 PAFZZ PAOZZ 5999-01-382-8223 30554 88-20188 . COVER, BATTERY TERMI 4
18 PAFZZ PAOZZ 5310-00-913-8881 30554 13218E0320-293 . NUT, PLAIN, HEX 4
19 PAFZZ PAOZZ 5310-01-257-7590 96906 MS51412-7 . WASHER, FLAT 4
20 XDFZZ XB 6160-01-480-5918 30554 88-21684 . BATTERY HOLD DOWN ASSY 1
21 PAFZZ PAOZZ 5306-01-366-4527 30554 88-21685 . BOLT, HOOK 2
22 PAFZA PAOZA 6140-01-446-9498 19207 6TMF/TYPEI . BATTERY, STORAGE 2
22 PAFZA PAOZA 6140-01-457-4339 30554 88-22800 . BATTERY 2
23 XDFZZ XB 6160-01-553-5219 30554 88-21683 . TRAY, BATTERY 1
24 PAFZZ PAOZZ 5305-00-269-3233 80204 B1821BH038F063N . SCREW, CAP, HEX HEAD 2
25 PAFZZ PAOZZ 5310-00-011-5093 80205 MS35338-65 . WASHER, LOCK .... 6 26 AFFFF AOO 6150-01-494-
3982 30554 88-22309 . CABLE ASSY, BATTERY (NOT
SHOWN) 1
27 PAFZZ PAOZZ 5940-01-369-2872 30554 88-22119-14 . . TERMINAL, LUG 1
28 PAFZZ PAOZZ 5940-00-804-0520 30554 88-22119-22 . . TERMINAL, LUG 1

(1)	(2)		(3)	(4)	(5)	(6)	(7)
	SMR CODE						
	a.	b.					
ITEM NO	ARMY	AIR FORCE	NSN	CAGEC	PART NUMBER	DESCRIPTION AND USABLE ON CODE (UOC)	QTY

29 MFFZZ MOO 30554 88-22309-3 . . INSULATION, SLEEVING MAKE FROM P/N M23053/5-110-9 (81349), AS REQUIRED 1

30 MFFZZ MOO 30554 88-22309-1 . . WIRE, ELECTRIC MAKE FROM P/N J1127SGR1AWGBLK (81343), 30.0 INCHES REQUIRED 1

31 AFFFF AOO 6150-01-494-3985 30554 88-22311 . CABLE ASSY, BATTERY (NOT SHOWN) 1

32 PAFZZ PAOZZ 5940-01-369-2872 30554 88-22119-14 . . TERMINAL, LUG 1

33 PAFZZ PAOZZ 5940-01-369-2268 30554 88-22119-11 . . TERMINAL, LUG .... 1 34 MFFZZ MOO 30554 88-22311-4 . . INSULATION, SLEEVING MAKE FROM P/N M23053/5-110-2 (81349), AS REQUIRED 2

35 MFFZZ MOO 30554 88-22311-5 . . INSULATION, SLEEVING MAKE FROM P/N M23053/5-110-9 (81349), AS REQUIRED 1

36 MFFZZ MOO 30554 88-22311-1 . . WIRE, ELECTRIC MAKE FROM P/N J1127SGR1AWGBLK (81343), 29.0 INCHES REQUIRED 1

37 PAFZZ PAOZZ 5310-01-012-3595 30554 69-561-6 . NUT, PLAIN, ASSEMBLED 4

38 PAFZZ PAOZZ 5306-01-156-7663 30554 88-20260-21 . SCREW, HEX WASHERHEAD 4

39 PAFZZ PAOZZ 5935-01-044-8382 19207 11682345 . CONNECTOR, RECEPTACLE 1

END OF FIGURE

FIELD AND SUSTAINMENT MAINTENANCE

15 kW 50/60 AND 400 Hz SKID MOUNTED TACTICAL QUIET GENERATOR SETS
GROUP 02 HOUSING: ACCESS DOORS AND COVERS

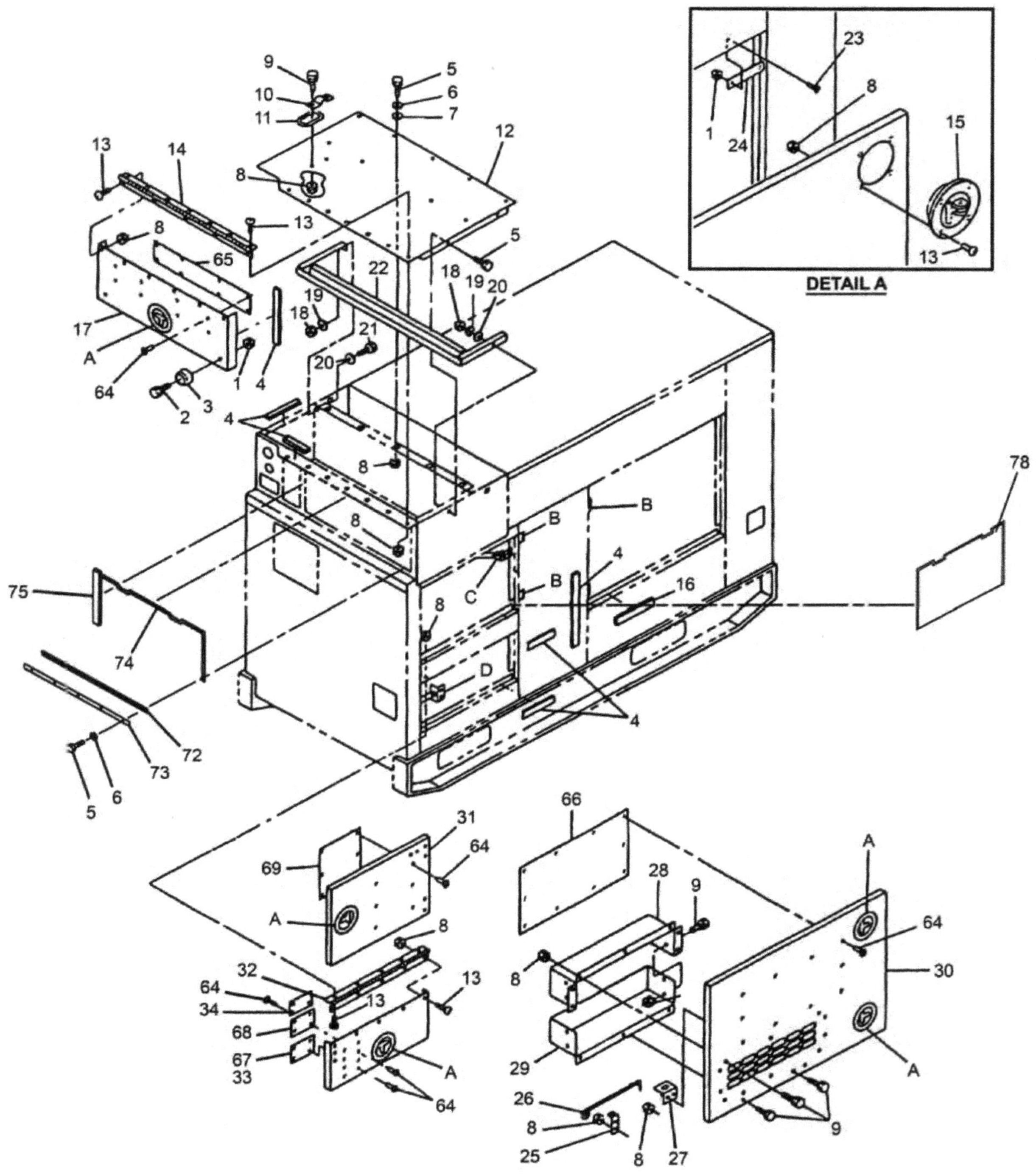

Figure 2. Access Doors and Covers (Sheet 1 of 5).

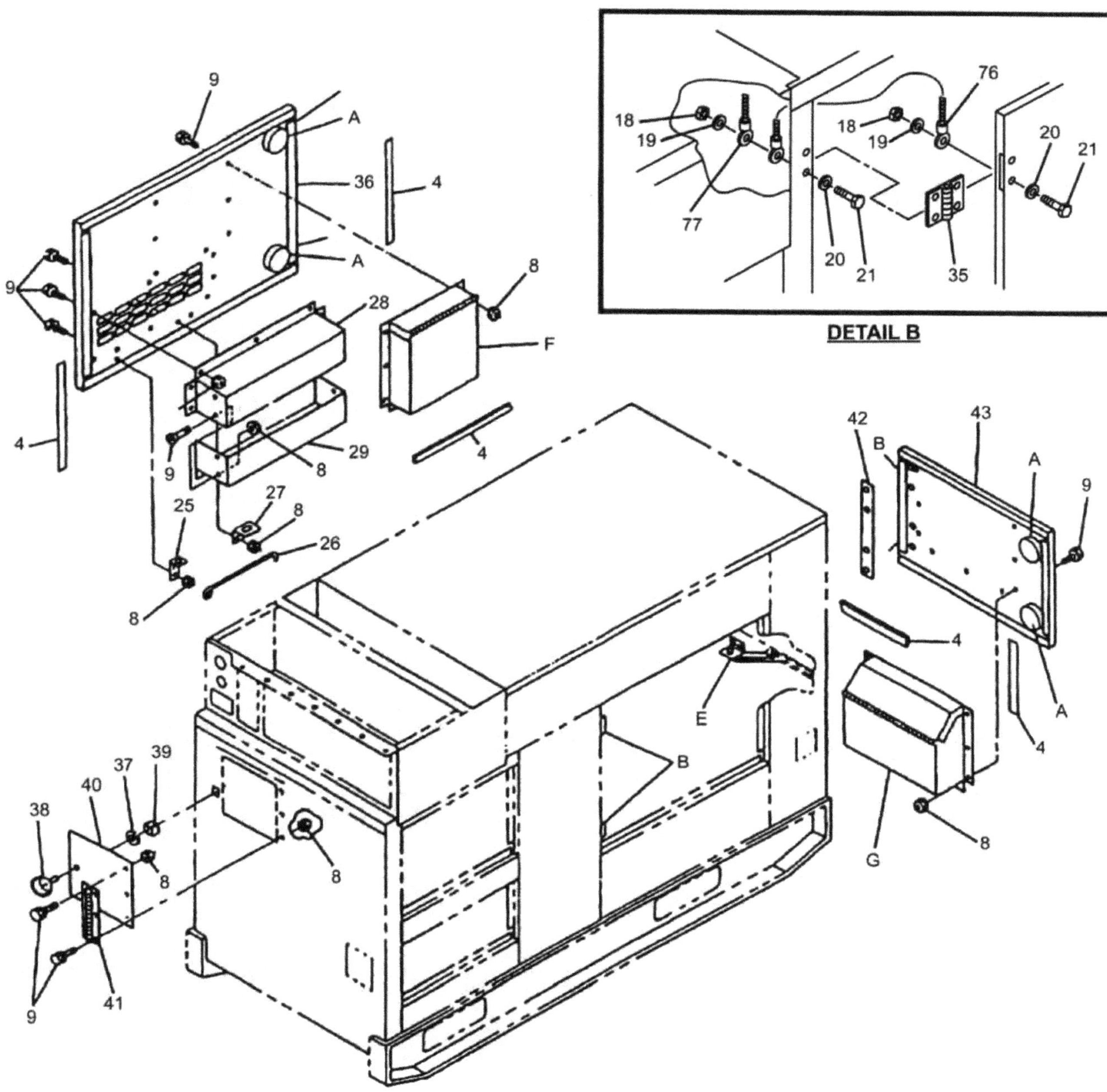

Figure 2. Access Doors and Covers (Sheet 2 of 5).

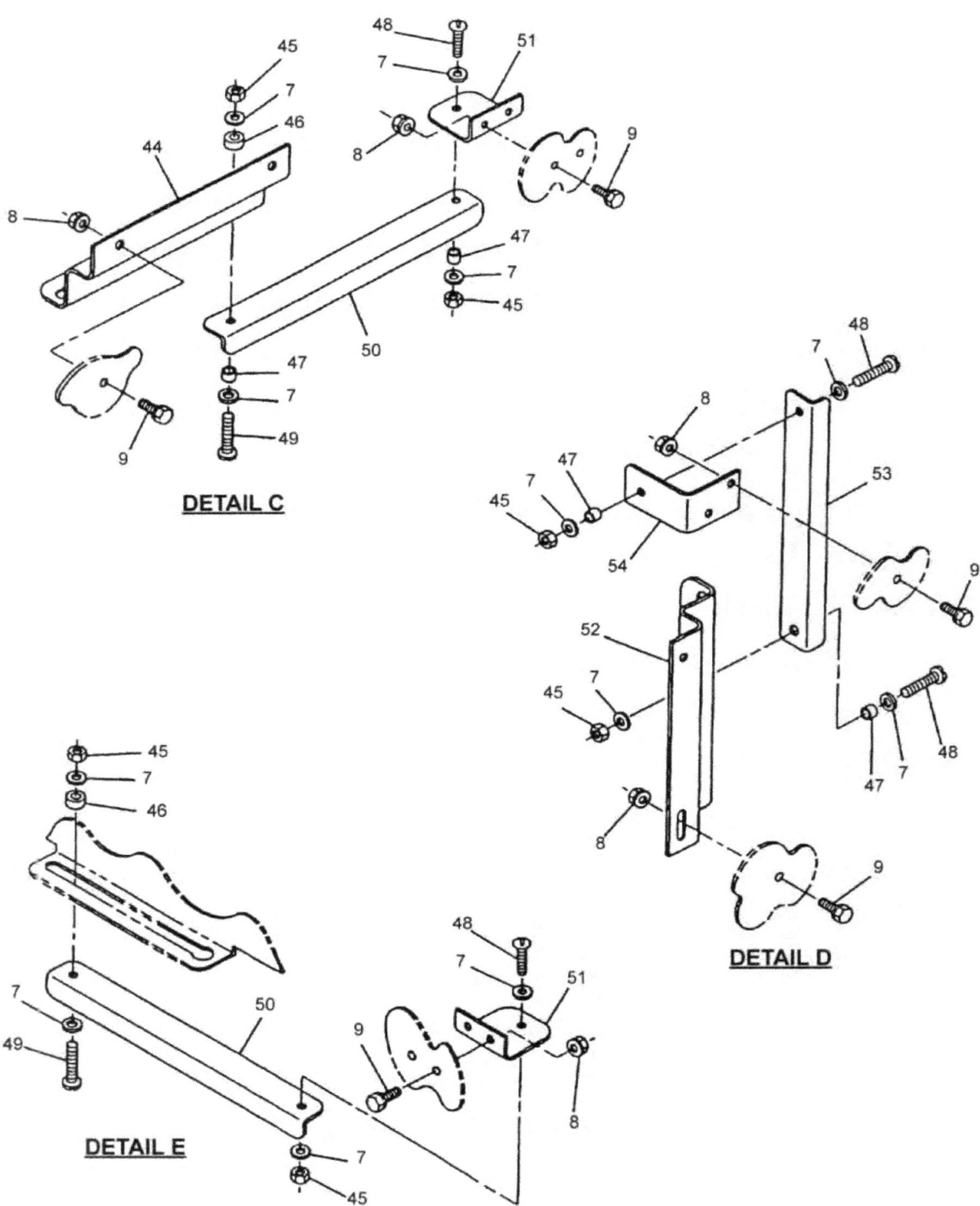

Figure 2. Access Doors and Covers (Sheet 3 of 5).

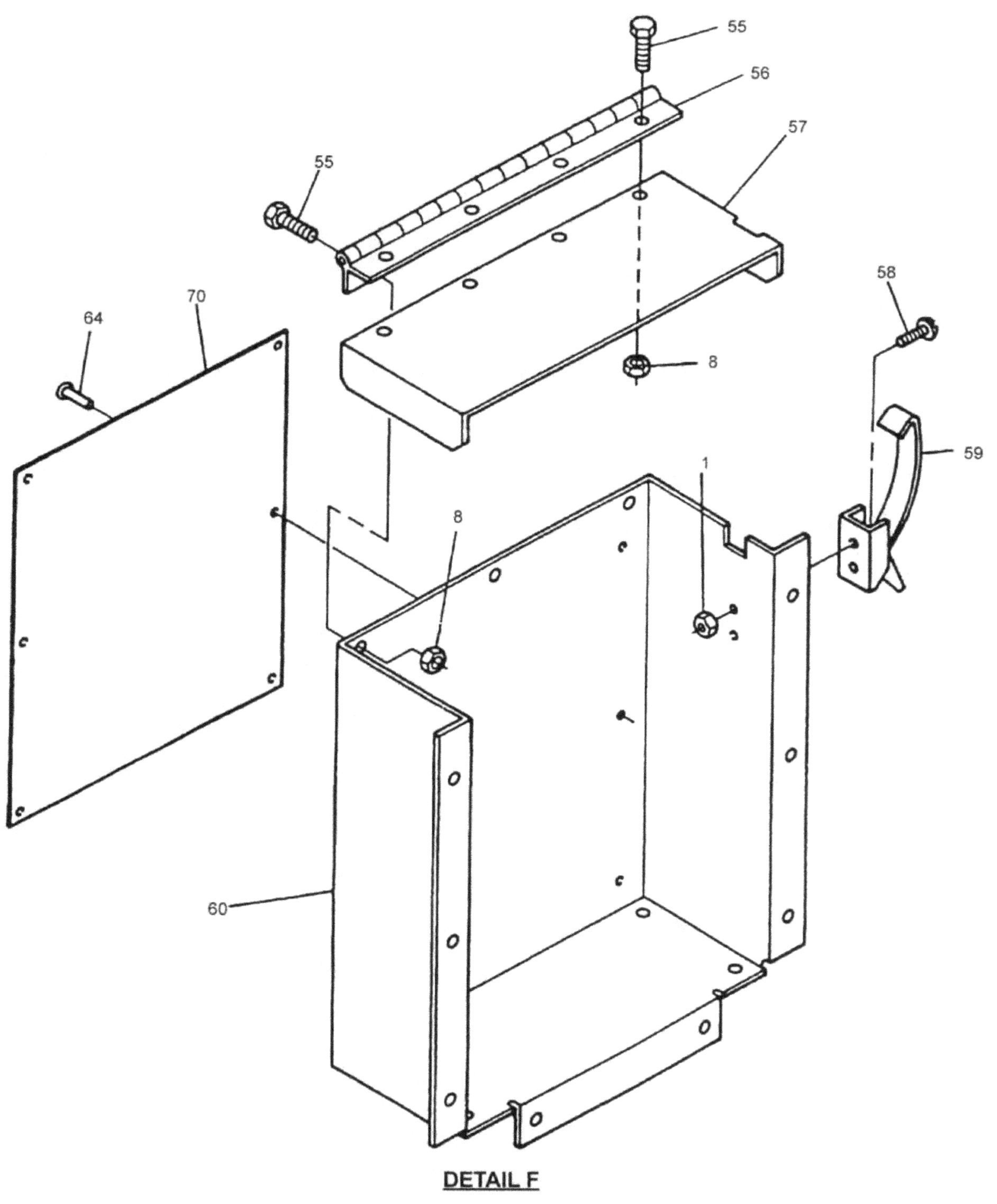

Figure 2. Access Doors and Covers (Sheet 4 of 5).

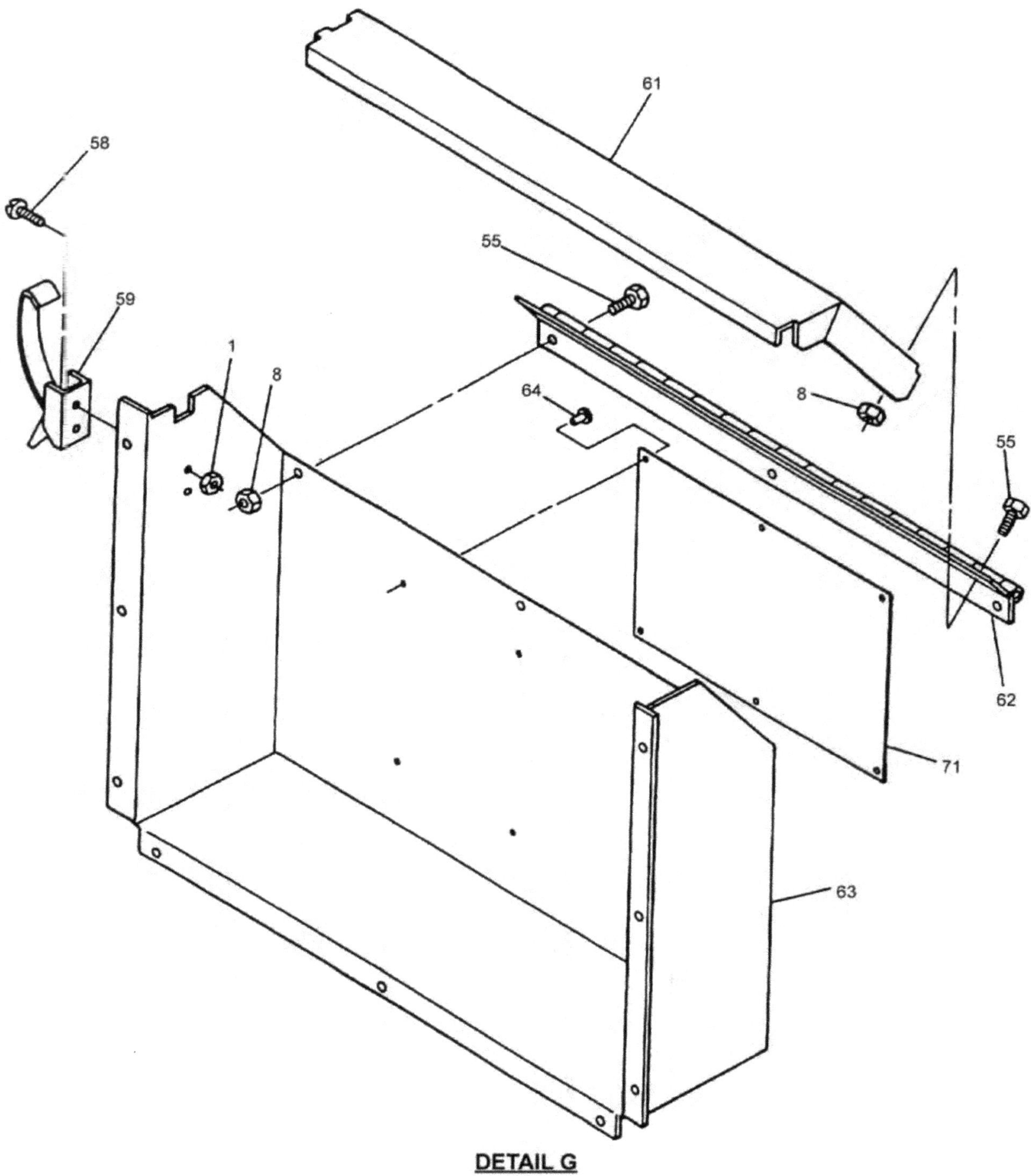

DETAIL G

Figure 2. Access Doors and Covers (Sheet 5 of 5).

(1)	(2)		(3)	(4)	(5)	(6)	(7)
	SMR CODE						
	a.	b.					
ITEM NO	ARMY	AIR FORCE	NSN	CAGEC	PART NUMBER	DESCRIPTION AND USABLE ON CODE (UOC)	QTY

GROUP 02 HOUSING

FIG. 2 ACCESS DOORS AND COVERS

1 PAFZZ PAOZZ 5310-00-063-7360 30554 69-561-2 . NUT, PLAIN, ASSEMBLED 22
2 PAFZZ PAOZZ 5305-00-036-6972 30554 69-662-20 . SCREW, ASSEMBLED 2
3 PAFZZ PAOZZ 5340-01-532-0643 30554 88-20278 . BUMPER, DOOR .... 2 4 MFFZZ MOO 30554 88-21604-295 . SEAL, DOOR MAKE FROM P/N
88-22705 (30554), AS REQUIRED
UOC: EMK 1
4 MFFZZ MOO 30554 97-24604-295 . SEAL, DOOR MAKE FROM P/N
88-22705 (30554), AS REQUIRED
UOC: 86Q.................................. 1
4 MFFZZ MOO 30554 88-21605-295 . SEAL, DOOR MAKE FROM P/N
88-22705 (30554), AS REQUIRED
UOC: YNN 1
4 MFFZZ MOO 30554 97-24605-295 . SEAL, DOOR MAKE FROM P/N
88-22705 (30554), AS REQUIRED
UOC: 86R 1
5 PAFZZ PAOZZ 5305-01-365-6313 30554 88-20260-23 . SCREW, CAP, HEX HEAD 16
6 PAFZZ PAOZZ 5310-00-274-8710 80205 MS35338-62 . WASHER, LOCK 12
7 PAFZZ PAOZZ 5310-01-234-9416 96906 MS51412-2 . WASHER, FLAT 24
8 PAFZZ PAOZZ 5310-01-012-3595 30554 69-561-6 . NUT, PLAIN, ASSEMBLED 115
9 PAFZZ PAOZZ 5306-01-156-7663 30554 88-20260-21 . SCREW, HEX WASHERHEAD 72
10 XDFZZ XB 30554 88-20277 . RETAINER, DOOR RING 1
11 XDFZZ XB 30554 88-20276 . RING, DOOR 1
12 XDFZZ XB 30554 88-21870 . TOP, CONTROL BOX 1
13 PAFZZ PAOZZ 5305-01-464-6667 30554 88-22793-4 . SCREW, MACHINE .... 58 14 MFFZZ MOO 30554 88-21877 . HINGE, DOOR, CONTROL MAKE
FROM P/N AA07060408 (03007), 27.75
INCHES REQD 1
15 PAFZZ PAOZZ 5340-01-467-0760 30554 88-21099 . LATCH, TEE HANDLE .... 9 16 MFFZZ MOO 30554 88-21604-356 . SEAL, DOOR MAKE FROM P/N
88-22708 (30554), 74.8 INCHES
REQUIRED
UOC: EMK 1
16 MFFZZ MOO 30554 97-24604-356 . SEAL, DOOR MAKE FROM P/N
88-22708 (30554), 74.8 INCHES
REQUIRED
UOC: 86Q.................................. 1
16 MFFZZ MOO 30554 88-21605-356 . SEAL, DOOR MAKE FROM P/N
88-22708 (30554), 74.8 INCHES
REQUIRED
UOC: YNN 1
16 MFFZZ MOO 30554 97-24605-356 . SEAL, DOOR MAKE FROM P/N
88-22708 (30554), 74.8 INCHES
REQUIRED
UOC: 86R 1
17 XDFZZ XB 5340-01-528-9207 30554 88-21875 . DOOR, CONTROL BOX 1
18 PAFZZ PAOZZ 5310-01-466-6312 30554 88-22790-1 . NUT, PLAIN, HEX 34
19 PAFZZ PAOZZ 5310-00-274-8715 80205 MS35338-63 . WASHER, LOCK 34
20 PAFZZ PAOZZ 5310-01-103-6042 96906 MS51412-4 . WASHER, FLAT 34
21 PAFZZ PAOZZ 5305-01-531-4346 30554 88-20260-31 . SCREW, HEX WASHERHEAD 34

(1) ITEM NO	(2) SMR CODE a. ARMY	b. AIR FORCE	(3) NSN	(4) CAGEC	(5) PART NUMBER	(6) DESCRIPTION AND USABLE ON CODE (UOC)	(7) QTY
22	XDFZZ	XB		30554	88-22077	. STIFFENER ASSY, CON	1
23	PAFZZ	PAOZZ	5305-01-466-9756	30554	88-22791-1	. SCREW, MACHINE	16
24	XDFZZ	XB		30554	88-20123	. PLATE, STRIKER	8
25	XDFZZ	XB		30554	88-22501	. BRACKET, HOLDING ROD	2
26	XDFZZ	XB	5340-01-533-0084	30554	88-22472	. ROD, DOOR HOLDING	2
27	XDFZZ	XB		30554	88-22500	. BRACKET, LATCH-HOLD	2
28	XDFZZ	XB		30554	88-22088	. BAFFLE, UPPER	2
29	XDFZZ	XB		30554	88-22089	. BAFFLE, LOWER	2
30	XDFZZ	XB		30554	88-22070	. DOOR, RS	1
31	XDFZZ	XB		30554	88-21585	. DOOR, OUTPUT BOX UOC: EMK, YNN	1
31	XDFZZ	XB		30554	97-24138	. DOOR, OUTPUT BOX UOC: 86Q, 86R	1
32	MFFZZ	MOO		30554	88-21820	. HINGE, DOOR, LOAD MAKE FROM P/N AA06060406 (03007), 21.26 INCHES REQUIRED	1
33	XDFZZ	XB		30554	88-21771	. DOOR, ACCESS LOAD TER	1
34	MFFZZ	MDO		30554	88-22738	. PLATE, IDENTIFICATION, SWTCH BOX RECEPTACLE	1
35	PAFZZ	PAOZZ	5340-01-368-6048	30554	88-21098	. HINGE, BUTT	8
36	XDFZZ	XB	5340-01-392-8822	30554	88-21890	. DOOR, ACCESS, LS	1
37	PAFZZ	PAOZZ	5310-00-822-8525	30554	88-20052	. RETAINER	1
38	PAFZZ	PAOZZ	5307-01-374-4451	30554	88-22626	. STUD, 1/4 TURN	1
39	PAFZZ	PAOZZ	5325-01-301-7903	30554	88-20050	. RECEPTACLE, TURNLOCK	1
40	XDFZZ	XB		30554	88-21732	. DOOR, ACCESS, AIR CL	1
41	MFFZZ	MOO		30554	88-21733	. HINGE, DOOR, AIR CL MAKE FROM P/N AA06060406 (03007), 8.75 INCHES REQUIRED. REMOVE CENTER PIN, REVERSE	1
42	XDFZZ	XB		30554	88-22666	. PLATE, SPACER, HINGE	1
43	XDFZZ	XB		30554	88-22063	. DOOR, BATTERY COMPA	1
44	XDFZZ	XB		30554	88-22478	. BRACKET, DOOR LINK	1
45	PAFZZ	PAOZZ	5310-01-406-1672	30554	88-21930-4	. NUT, SELF-LOCKING, HEX	6
46	XDFZZ	XB	5365-01-431-4540	30554	88-22483	. SPACER, SLEEVE	2
47	XDFZZ	XB	5365-01-431-4603	30554	88-22482	. SPACER, RING	4
48	PAFZZ	PAOZZ	5305-01-334-8683	96906	MS51493-2	. SCREW, MACHINE	4
49	PAFZZ	PAOZZ	5305-01-406-1192	96906	MS51493-3	. SCREW, MACHINE	2
50	XDFZZ	XB		30554	88-20461	. LINK, DOOR SUPPORT	2
51	XDFZZ	XB		30554	88-22479	. SUPPORT, DOOR LINK	2
52	XDFZZ	XB		30554	88-22474	. BRACKET, DOOR SUPPORT	1
53	XDFZZ	XB		30554	88-22476	. LINK, DOOR SUPPORT	1
54	XDFZZ	XB		30554	88-22475	. ANGLE, DOOR SUPPORT	1
55	PAFZZ	PAOZZ	5306-01-368-8041	30554	88-20260-20	. SCREW, HEX WASHERHEAD	14
56	MFFZZ	MOO	5340-01-531-7746	30554	88-20014	. HINGE, DOCUMENT BOX	1
57	XDFZZ	XB		30554	88-20013	. LID, DOCUMENT BOX	1
58	PAFZZ	PAOZZ	5305-00-211-9344	30554	69-662-17	. SCREW, ASSEMBLED	4
59	PAFZZ	PAOZZ	5340-01-396-0454	30554	88-20252	. LATCH, DRAW	2
60	XDFZZ	XB		30554	88-21998	. BOX, DOCUMENT	1
61	XDFZZ	XB		30554	88-21814	. LID, STORAGE BOX	1
62	MFFZZ	MOO	5340-01-535-0248	30554	88-21815	. HINGE, STORAGE BOX	1
63	XDFZZ	XB		30554	88-21813	. BOX, STORAGE	1
64	PAFZZ	PAOZZ	5320-01-019-5694	81349	M24243/1A402	. RIVET, BLIND	46

(1) ITEM NO	(2) SMR CODE a. ARMY	b. AIR FORCE	(3) NSN	(4) CAGEC	(5) PART NUMBER	(6) DESCRIPTION AND USABLE ON CODE (UOC)	(7) QTY
65	MDFZZ	MDO		30554	88-21635	. PLATE, IDENTIFICATION, OPERATING INSTRUCTIONS UOC: EMK	1
65	MDFZZ	MDO		30554	88-22078	. PLATE, IDENTIFICATION, OPERATING INSTRUCTIONS UOC: YNN	1
65	MDFZZ	MDO		30554	97-24053	. PLATE, IDENTIFICATION, OPERATING INSTRUCTIONS UOC: 86Q, 86R	1
66	MDFZZ	MDO		30554	88-22211	. PLATE, IDENTIFICATION, WIRING DIAGRAM UOC: EMK, YNN	1
66	MDFZZ	MDO		30554	97-24052	. PLATE, IDENTIFICATION, WIRING DIAGRAM UOC: 86Q, 86R	1
67	MDFZZ	MDO		30554	88-20110	. PLATE, IDENTIFICATION, CAUTION, VOLTAGE CONNECTION	1
68	MDFZZ	MDO		30554	88-20126	. PLATE, IDENTIFICATION, GROUNDING STUD	1
69	MDFZZ	MDO		30554	88-22210	. PLATE, IDENTIFICATION, SCHEMATIC DIAGRAM UOC: EMK, YNN	1
69	MDFZZ	MDO		30554	97-24051	. PLATE, IDENTIFICATION, SCHEMATIC DIAGRAM UOC: 86Q, 86R	1
70	MDFZZ	MDO		30554	88-22480	. PLATE, IDENTIFICATION, FUEL SYSTEM DIAGRAM	1
71	MDFZZ	MDO		30554	88-21603	. PLATE, IDENTIFICATION, INSTRUCTION, BATTERY CONNECTION	1
72	PAFZZ	PAOZZ	5999-01-574-4372	30554	97-24124	. SEAL, EMI UOC: 86Q, 86R	1
73	XDFZZ	XB		30554	97-24130	. BRACKET, RETAINING UOC: 86Q, 86R	1
74	PAFZZ	PAOZZ	5330-01-572-3123	30554	97-24137	. SEAL, EMI UOC: 86Q, 86R	1
75	PAFZZ	PAOZZ	5330-01-572-3123	30554	97-24131	. SEAL, EMI UOC: 86Q, 86R	1
76	PAFZZ	PAOZZ	5999-01-572-0977	30554	97-24128	. GROUND STRAP UOC: 86Q, 86R	2
77	PAFZZ	PAOZZ	5999-01-572-0957	30554	97-24129	. GROUND STRAP UOC: 86Q, 86R	2
78	PAFZZ	PAOZZ	5330-01-572-6232	30554	97-24132	. SEAL, EMI UOC: 86Q, 86R	AR

END OF FIGURE

FIELD AND SUSTAINMENT MAINTENANCE

15 kW 50/60 AND 400 Hz SKID MOUNTED TACTICAL QUIET GENERATOR SETS

GROUP 02 HOUSING: FRONT HOUSING PANELS

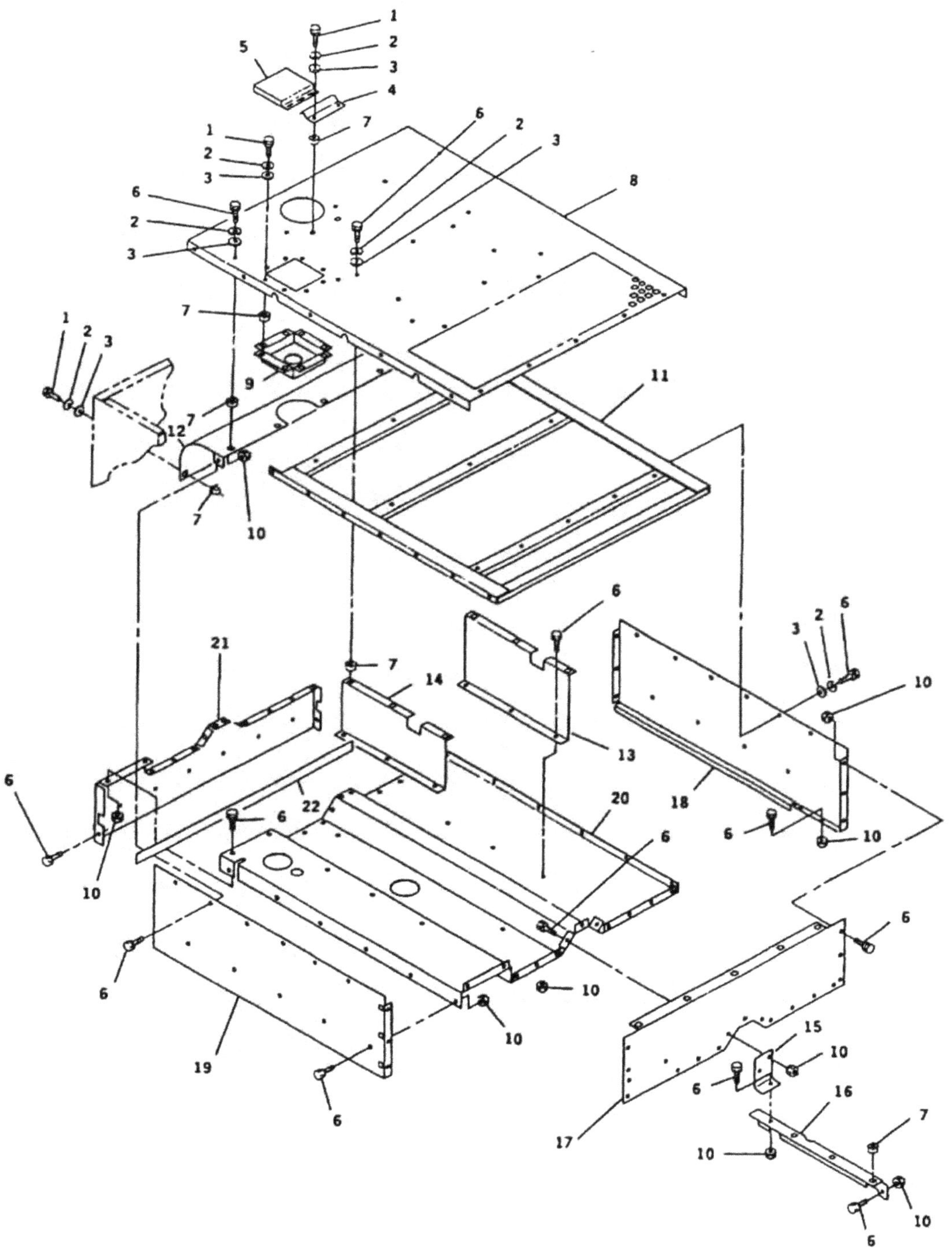

Figure 3. Front Housing Panels (Sheet 1 of 2).

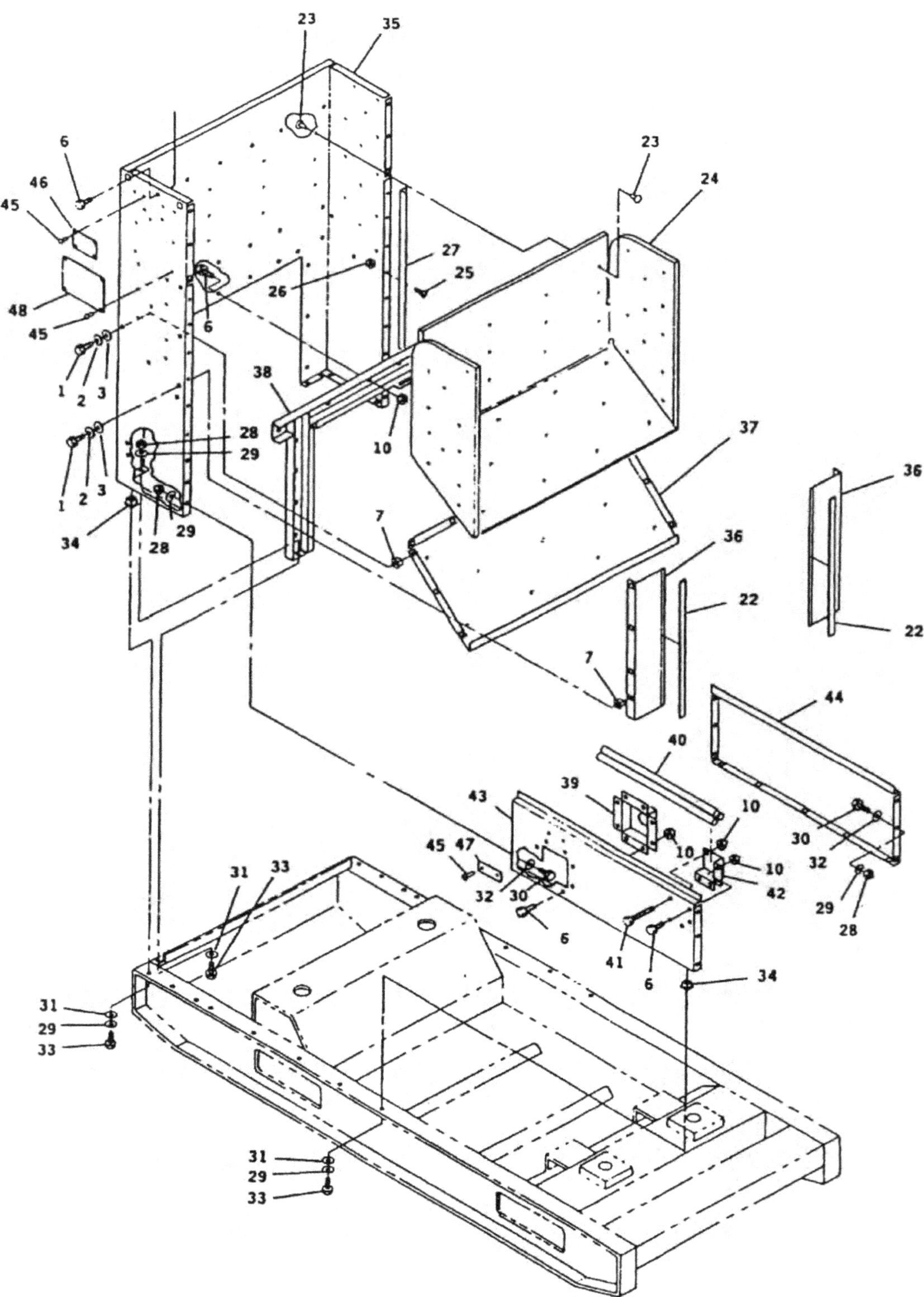

Figure 3. Front Housing Panels (Sheet 2 of 2).

(1) (2) (3) (4) (5) (6) (7)
SMR CODE
a. b.
ITEM NO ARMY AIR FORCE NSN CAGEC PART NUMBER DESCRIPTION AND USABLE ON CODE (UOC) QTY

GROUP 02 HOUSING

FIG. 3 FRONT HOUSING PANELS

1 PAFZZ PAOZZ 5305-01-365-6313 30554 88-20260-23 . SCREW, CAP, HEX HEAD 16
2 PAFZZ PAOZZ 5310-00-274-8710 80205 MS35338-62 . WASHER, LOCK 56
3 PAFZZ PAOZZ 5310-01-234-9416 96906 MS51412-2 . WASHER, FLAT 56
4 XDFZZ XB 5340-01-367-8956 30554 88-22430 . BRACKET, MOUNTING 1
5 XDFZZ XB 2990-01-389-3003 30554 88-22429 . COVER, EXHAUST 1
6 PAFZZ PAOZZ 5306-01-156-7663 30554 88-20260-21 . SCREW, HEX WASHERHEAD 138
7 PAFZZ PAOZZ 5310-00-903-8595 30554 88-21674-2 . NUT, CAGE 46
8 XDFZZ XB 30554 88-21598 . HOUSING, TOP 1
9 XDFZZ XB 30554 88-21988 . PANEL, FILL RADIATOR 1
10 PAFZZ PAOZZ 5310-01-012-3595 30554 69-561-6 . NUT, PLAIN, ASSEMBLED 106
11 XDFZZ XB 30554 88-22116 . STIFFENER, ROOF 1
12 XDFZZ XB 30554 88-21962 . DEFLECTOR, AIR UPPER 1
13 XDFZZ XB 30554 88-21600 . CHANNEL, DUCT RIGHT SIDE 1
14 XDFZZ XB 30554 88-22513 . CHANNEL, DUCT LEFT SIDE 1
15 XDFZZ XB 30554 88-22528 . BRACKET, SUPPORT OU 1
16 XDFZZ XB 30554 88-22702 . ANGLE, SUPPORT 1
17 XDFZZ XB 30554 88-22510 . PANEL, DUCT FLOOR 1
18 XDFZZ XB 30554 88-21928 . PANEL, TOP RIGHT SIDE 1
19 XDFZZ XB 30554 88-21596 . PANEL, TOP LEFT SIDE 1
20 XDFZZ XB 30554 88-21601 . FLOOR, DUCT 1
21 XDFZZ XB 30554 88-22512 . PANEL, FRONT FLOOR .... 1 22 MFFZZ MOO 30554 88-21604-317 . SEAL, RUBBER MAKE FROM P/N
MIL-R-6130 TY II GRA (81349), AS REQUIRED
UOC: EMK 1
22 MFFZZ MOO 30554 97-24604-317 . SEAL, RUBBER MAKE FROM P/N
MIL-R-6130 TY II GRA (81349), AS REQUIRED
UOC: 86Q.................................. 1
22 MFFZZ MOO 30554 88-21605-317 . SEAL, RUBBER MAKE FROM P/N
MIL-R-6130 TY II GRA (81349), AS REQUIRED
UOC: YNN 1
22 MFFZZ MOO 30554 97-24605-317 . SEAL, RUBBER MAKE FROM P/N
MIL-R-6130 TY II GRA (81349), AS REQUIRED
UOC: 86R 1
23 PAFZZ PAOZZ 5320-01-366-4394 30554 88-20210 . CLIP, PANEL .... 92 24 MFFZZ MOO 30554 88-22595 . INSULATION, HOUSING, FR MAKE
FROM P/N FF40JM02 (28818), 51 INCHES X 36 1
25 PAFZZ PAOZZ 5305-00-059-4568 80205 MS35190-253 . SCREW, MACHINE 12
26 PAFZZ PAOZZ 5310-01-396-5836 30554 88-21930-3 . NUT, SELF-LOCKING 12
27 XDFZZ XB 30554 88-22667 . CHANNEL 2
28 PAFZZ PAOZZ 5310-01-466-6312 30554 88-22790-1 . NUT, PLAIN, HEX 14
29 PAFZZ PAOZZ 5310-00-274-8715 80205 MS35338-63 . WASHER, LOCK 34
30 PAFZZ PAOZZ 5305-01-531-4346 30554 88-20260-31 . SCREW, HEX WASHERHEAD 12
31 PAFZZ PAOZZ 5310-01-234-9415 96906 MS51412-5 . WASHER, FLAT 18
32 PAFZZ PAOZZ 5310-01-103-6042 96906 MS51412-4 . WASHER, FLAT 16

(1)	(2)		(3)	(4)	(5)	(6)	(7)
	SMR CODE						
	a.	b.					
ITEM NO	ARMY	AIR FORCE	NSN	CAGEC	PART NUMBER	DESCRIPTION AND USABLE ON CODE (UOC)	QTY

33 PAFZZ PAOZZ 5306-01-366-7075 30554 88-20260-33 . SCREW, HEX WASHERHEAD 22
34 PAFZZ PAOZZ 5340-00-297-0312 30554 88-21674-3 . NUT, CAGE 20
35 XDFZZ XB 30554 88-21913 . HOUSING, FRONT 1
36 XDFZZ XB 30554 88-21694 . PANEL, RADIATOR 2
37 XDFZZ XB 30554 88-22051 . PANEL, AIR DEFLECTOR 1
38 XDFZZ XB 30554 88-22514 . CHANNEL, SUPPORT 1
39 XDFZZ XB 30554 88-21952 . BOX, SLAVE RECEPTAC 1
40 XDFZZ XB 5975-00-878-3791 58536 AA55804-3B . ROD, GROUND 9FT 1
41 PAFZZ PAOZZ 5305-00-071-2236 80205 MS90725-15 . SCREW, CAP, HEX HEAD 2
42 XDFZZ XB 30554 88-21889 . SUPPORT, GROUND ROD 2
43 XDFZZ XB 30554 88-21590 . SILL, DOOR LEFT SIDE 1
44 XDFZZ XB 30554 88-22032 . SILL, DOOR RIGHT SIDE 1
45 PAFZZ PAOZZ 5320-01-019-5694 81349 M24243/1A402 . RIVET, BLIND .... 10 46 MDFZZ MDO 30554 88-20063-05 . PLATE, IDENTIFICATION UOC: EMK 1

46 MDFZZ MDO 30554 88-20063-06 . PLATE, IDENTIFICATION UOC: YNN 1

46 MDFZZ MDO 30554 97-24050-01 . PLATE, IDENTIFICATION UOC: 86Q.................................. 1

46 MDFZZ MDO 30554 97-24050-02 . PLATE, IDENTIFICATION UOC: 86R 1

47 MDFZZ MDO 30554 88-20075 . PLATE, IDENTIFICATION, SLAVE RECEPTACLE 1

48 MDFZZ MDO 30554 88-21821-01 . PLATE, IDENTIFICATION, SET RATING UOC: EMK, 86Q............................ 1

48 MDFZZ MDO 30554 88-21821-02 . PLATE, IDENTIFICATION, SET RATING UOC: YNN, 86R............................ 1

END OF FIGURE

FIELD AND SUSTAINMENT MAINTENANCE

15 kW 50/60 AND 400 Hz SKID MOUNTED TACTICAL QUIET GENERATOR SETS
GROUP 02 HOUSING: REAR HOUSING PANELS

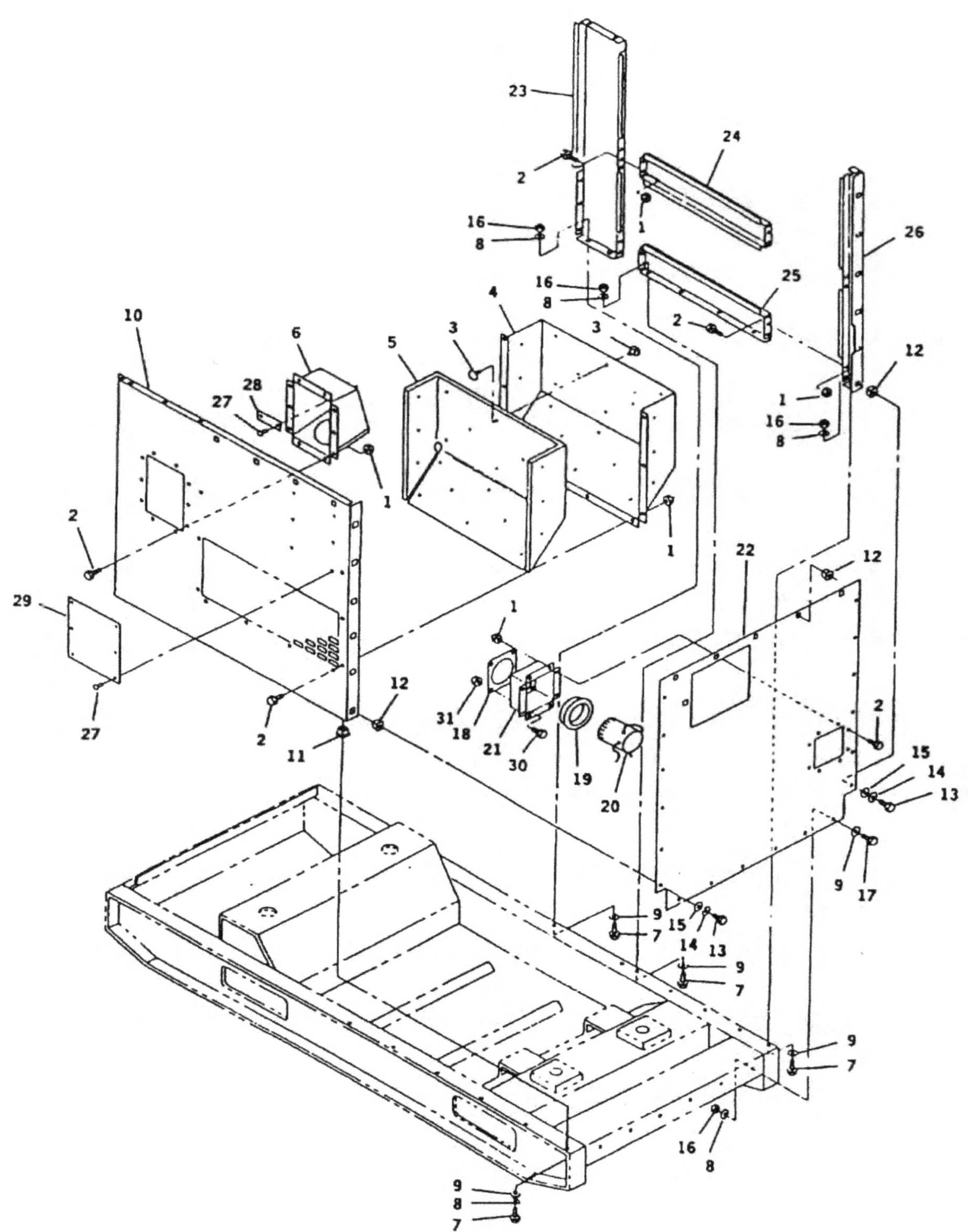

Figure 4. Rear Housing Panels.

(1) ITEM NO	(2) SMR CODE a. ARMY	b. AIR FORCE	(3) NSN	(4) CAGEC	(5) PART NUMBER	(6) DESCRIPTION AND USABLE ON CODE (UOC)	(7) QTY
						GROUP 02 HOUSING	
						FIG. 4 REAR HOUSING PANELS	
1	PAFZZ	PAOZZ	5310-01-012-3595	30554	69-561-6	. NUT, PLAIN, ASSEMBLED	31
2	PAFZZ	PAOZZ	5306-01-156-7663	30554	88-20260-21	. SCREW, HEX WASHERHEAD	31
3	PAFZZ	PAOZZ	5320-01-366-4394	30554	88-20210	. CLIP, PANEL	20
4	XDFZZ	XB		30554	88-22072	. BAFFLE, AIR INTAKE	1
5	MFFZZ	MOO		30554	88-22594	. INSULATION MAKE FROM P/N FF40JM02 (28818), 16 INCHES X 22 INCHES REQUIRED	1
6	XDFZZ	XB		30554	88-22661	. PANEL, FILLER NECK	1
7	PAFZZ	PAOZZ	5306-01-366-7075	30554	88-20260-33	. SCREW, HEX WASHERHEAD	10
8	PAFZZ	PAOZZ	5310-00-274-8715	80205	MS35338-63	. WASHER ,LOCK	20
9	PAFZZ	PAOZZ		96906	MS52412-5	. WASHER, FLAT	20
10	XDFZZ	XB		30554	88-21925	. PANEL, REAR LEFT SIDE	1
11	PAFZZ	PAOZZ	5340-00-297-0312	30554	88-21674-3	. NUT, CAGE	4
12	PAFZZ	PAOZZ	5310-00-903-8595	78553	88-21674-2	. NUT, CAGE	21
13	PAFZZ	PAOZZ	5305-01-365-6313	30554	88-20260-23	. SCREW, HEX WASHERHEAD	12
14	PAFZZ	PAOZZ	5310-00-274-8710	80205	MS35338-62	. WASHER, LOCK	12
15	PAFZZ	PAOZZ	5310-01-234-9416	96906	MS51412-2	. WASHER, FLAT	12
16	PAFZZ	PAOZZ	5310-01-466-6312	30554	88-22790-1	. NUT, PLAIN, HEX	13
17	PAFZZ	PAOZZ	5305-01-056-1501	30554	88-20260-32	. SCREW, HEX WASHERHEAD	6
18	XDFZZ	XB	5340-01-157-9475	30554	69-583	. COVER, ACCESS	1
19	PAFZZ	PAOZZ	5975-00-257-8055	30554	69-570-3	. BUSHING, ELECTRICAL	1
20	XDFZZ	XB	2920-01-388-2776	30554	88-20218	. COVER, ELECTRICAL	1
21	XDFZZ	XB		30554	88-21964	. BOX, LOAD TERMINAL	1
22	XDFZZ	XB		30554	88-21977	. PANEL, REAR	1
23	XDFZZ	XB		30554	88-22074	. PANEL, RIGHT SIDE	1
24	XDFZZ	XB		30554	88-22040	. SILL, OUTPUT BOX	1
25	XDFZZ	XB	5340-01-528-8654	30554	88-21770	. SILL, LOAD TERMINAL	1
26	XDFZZ	XB		30554	88-22037	. POST, CORNER, REAR	1
27	PAFZZ	PAOZZ	5320-01-019-5694	81349	M24243/1A402	. RIVET, BLIND	8
28	MDFZZ	MDO	9905-01-377-5094	30554	88-20102	. PLATE, IDENTIFICATION, EXTERNAL FUEL SUPPLY	1
29	MDFZZ	MDO		30554	88-21723	. PLATE, IDENTIFICATION, LIFTING AND TIEDOWN DIAGRAM	1
30	PAFZZ	PAOZZ	5305-00-038-3103	30554	69-662-35	. SCREW, ASSEMBLED	4
31	PAFZZ	PAOZZ	5310-00-052-3632	97403	13214E3291-3	. NUT, PLAIN	4

END OF FIGURE

FIELD AND SUSTAINMENT MAINTENANCE

15 kW 50/60 AND 400 Hz SKID MOUNTED TACTICAL QUIET GENERATOR SETS
GROUP 03 CONTROL BOX ASSEMBLY: CONTROL BOX INSTALLATION

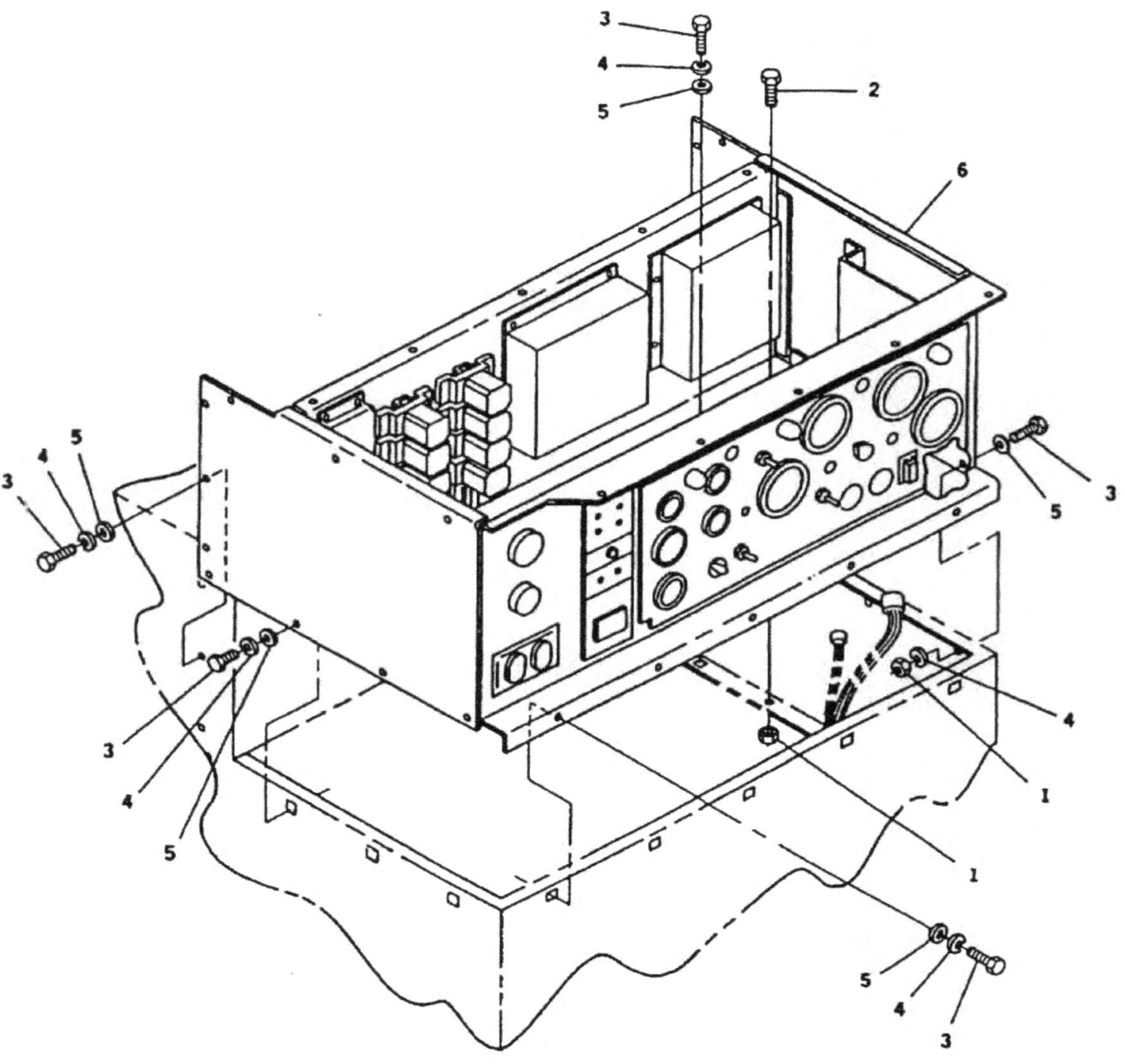

Figure 5. Control Box Installation.

(1) ITEM NO	(2) SMR CODE a. ARMY	b. AIR FORCE	(3) NSN	(4) CAGEC	(5) PART NUMBER	(6) DESCRIPTION AND USABLE ON CODE (UOC)	(7) QTY
						GROUP 03 CONTROL BOX ASSEMBLY	
						FIG. 5 CONTROL BOX INSTALLATION	
1	PAFZZ	PAOZZ	5310-01-012-3595	30554	69-561-6	NUT, PLAIN, ASSEMBLED	5
2	PAFZZ	PAOZZ	5306-01-156-7663	30554	88-20260-21	SCREW, HEX WASHERHEAD	5
3	PAFZZ	PAOZZ	5305-01-365-6313	30554	88-20260-23	SCREW, HEX WASHERHEAD	12
4	PAFZZ	PAOZZ	5310-00-274-8710	80205	MS35338-62	WASHER, LOCK	12
5	PAFZZ	PAOZZ	5310-01-234-9416	96906	MS51412-2	WASHER, FLAT	12
6	PBFFZ	PBFFZ	6110-01-397-2108	30554	88-22137	CONTROL BOX, ASSEMBLY UOC: EMK	1
6	PBFFZ	PBFFZ		30554	97-24080	CONTROL BOX, ASSEMBLY UOC: 86Q	1
6	PBFFZ	PBFFZ	6110-01-383-4531	30554	88-22138	CONTROL BOX, ASSEMBLY UOC: YNN	1
6	PBFFZ	PBFFZ		30554	97-24081	CONTROL BOX, ASSEMBLY UOC: 86R	1

END OF FIGURE

FIELD AND SUSTAINMENT MAINTENANCE

15 kW 50/60 AND 400 Hz SKID MOUNTED TACTICAL QUIET GENERATOR SETS
GROUP 03 CONTROL BOX ASSEMBLY: CONTROL PANEL SWITCHES/INDICATORS

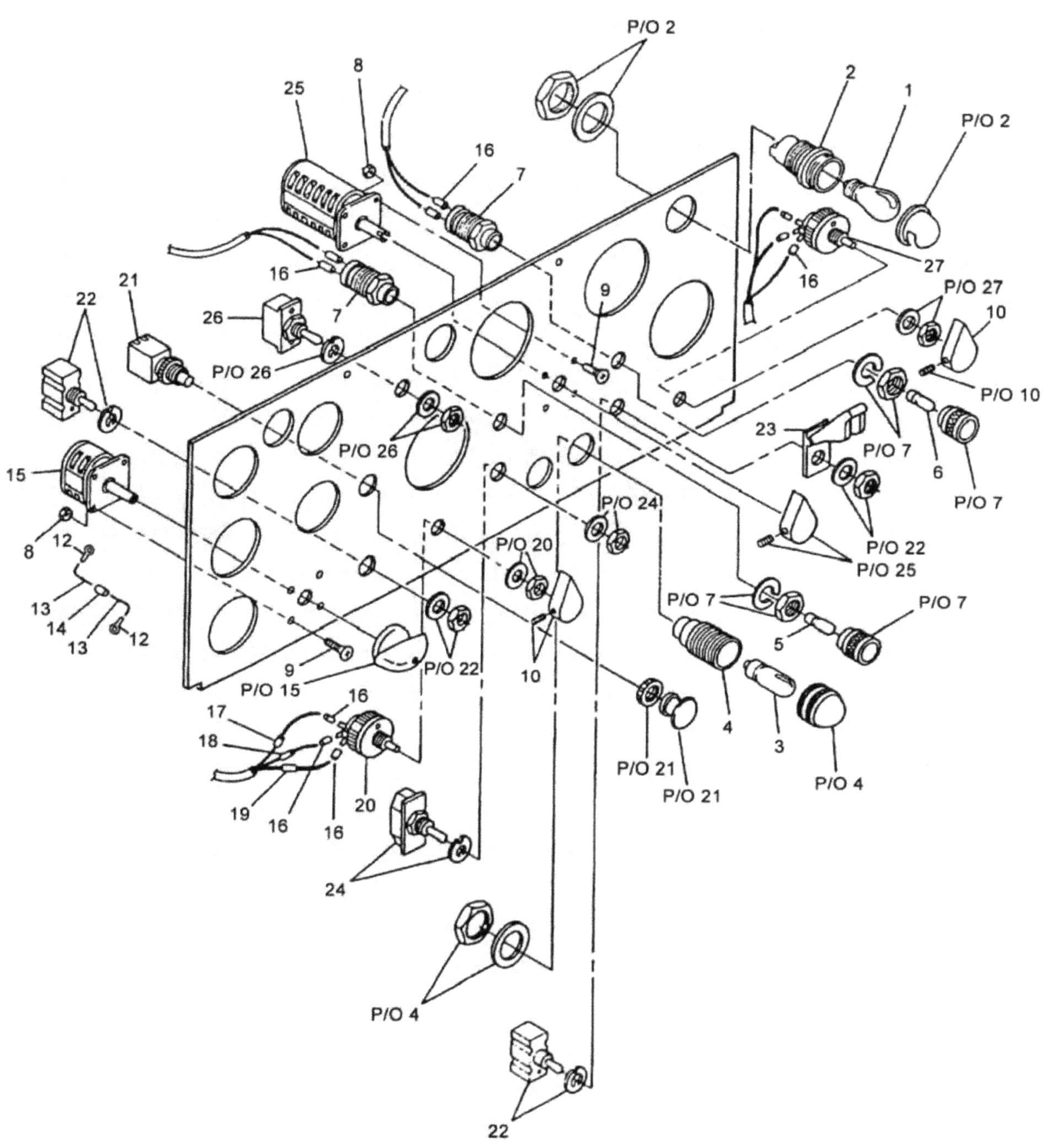

Figure 6. Control Panel Switches/Indicators (Sheet 1 of 2).

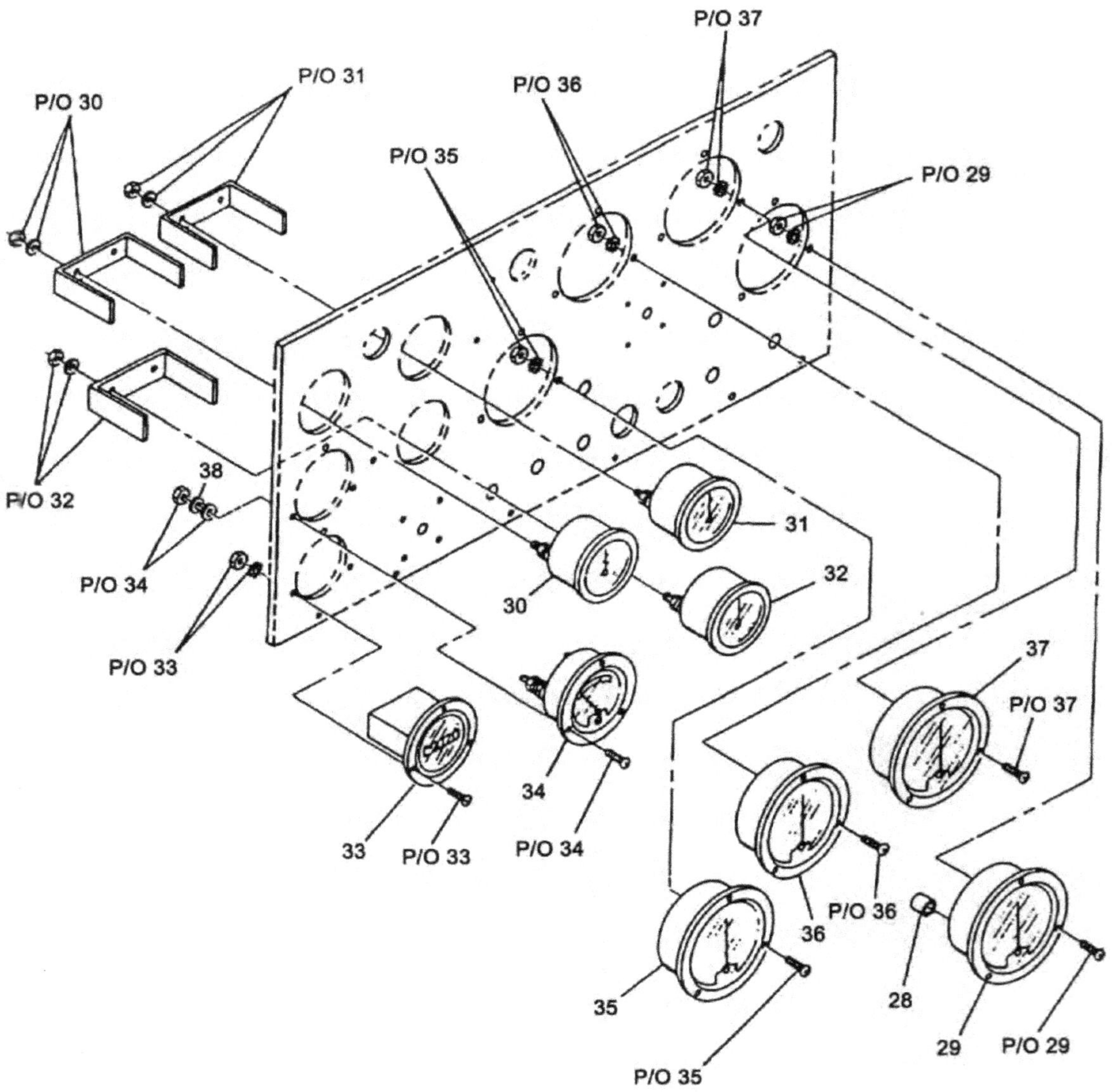

Figure 6. Control Panel Switches/Indicators (Sheet 2 of 2).

(1)	(2) SMR CODE a.	b.	(3)	(4)	(5)	(6)	(7)
ITEM NO	ARMY	AIR FORCE	NSN	CAGEC	PART NUMBER	DESCRIPTION AND USABLE ON CODE (UOC)	QTY

GROUP 03 CONTROL BOX ASSEMBLY

FIG. 6 CONTROL PANEL SWITCHES/INDICATORS

1 PAFZZ PAOZZ 6240-00-155-7878 30554 69-594 . . LAMP, INCANDESCENT GREEN 3
2 PAFZZ PAOZZ 6210-00-935-6919 30554 69-593 . . LIGHT, PANEL 3
3 PAFZZ PAOZZ 6240-01-466-3528 96906 A50452-1 . . LAMP, INCANDESCENT 2
4 PAFZZ PAOZZ 6210-00-831-8247 30554 88-21124 . . LIGHT, INDICATOR 2
5 PAFZZ PAOZZ 5980-01-076-8659 30554 88-21899 . . LAMP, LED, GREEN 1
6 PAFZZ PAOZZ 5980-01-198-6311 30554 88-21898 . . LAMP. LED, YELLOW 1
7 PAFZZ PAOZZ 6210-00-583-9349 96906 MS25041-5 . . LIGHT, INDICATOR 2
8 PAFZZ PAOZZ 5310-00-063-7360 30554 69-561-2 . . NUT, PLAIN, ASSEMBLED 8
9 PAFZZ PAOZZ 5305-00-071-1319 81348FFS92TYPE3 STYLE2C . . SCREW, MACHINE 8
10 PAFZZ PAOZZ 5355-00-559-8943 80205 MS91528-2K2B . . KNOB .... 2 11 AFFFF AOO 5961-01-470-4673 30554 88-22418-2 . . DIODE ASSEMBLY (NOT SHOWN) 1
12 PAFZZ PAOZZ 5940-00-143-4771 81343 MS25036-103 ...TERMINAL, LUG .... 2 13 MFFZZ MOO 30554 88-22418-2-2 ...SLEEVE, INSULATION MAKE FROM P/N M23053/5-104-0 (81349), AS REQUIRED 1
14 PAFZZ PAOZZ 5961-00-295-5757 50434 1901-0759 . . . SEMICONDUCTOR DEVICE 1
15 PAFZZ PAOZZ 5930-01-368-5160 30554 01-20002 . . SWITCH, ROTARY .... 1 16 MFFZZ MOO 30554 88-22137-105 . . INSULATION, SLEEVING MAKE FROM P/N M23053/5-105-0 (81349), AS REQUIRED 2
17 MFFZZ MOO 30554 88-22629-078 . . MARKER, IDENTIFICATION MAKE FROM P/N M23053/5-105-9 (81349), 1.25 INCHES REQUIRED 1
18 MFFZZ MOO 30554 88-22629-079 . . MARKER, IDENTIFICATION MAKE FROM P/N M23053/5-105-9 (81349), 1.25 INCHES REQUIRED 1
19 MFFZZ MOO 30554 88-22629-080 . . MARKER, IDENTIFICATION MAKE FROM P/N M23053/5-105-9 (81349), 1.25 INCHES REQUIRED 1
20 PAFZZ PAOZZ 5905-00-539-2573 30554 88-22773-1 . . RESISTOR, VARIABLE, ... 1 21 PAFZZ PAOZZ 5930-01-499-3684 30554 96-23737 . . SWITCH, ASSY, EMERGENCY STOP 1
22 PAFZZ PAOZZ 5930-01-368-2891 30554 88-21078 . . SWITCH, TOGGLE 2
23 PAFZZ PAOZZ 5930-00-615-6731 96906 MS25224-1 . . GUARD, SWITCH .... 1 24 PAFZZ PAOZZ 5930-01-368-2893 30554 88-20176 . . SWITCH, TOGGLE AC CIRCUIT INTERRUPTER 1
25 PAFZZ PAOZZ 5930-01-385-1894 30554 01-20003 . . SWITCH, ROTARY 1
26 PAFZZ PAOZZ 5930-01-366-0048 30554 88-22141 . . SWITCH, TOGGLE .... 1 27 PAFZZ PAOZZ 5905-00-556-5306 30554 88-22773-2 . . RESISTOR, VARIABLE, UOC: EMK, 86Q............................ 1
27 PAFZZ PAOZZ 5905-01-236-4041 30554 88-22773-3 . . RESISTOR, VARIABLE, UOC: YNN, 86R............................ 1
28 XDFZZ XB 5340-01-107-7559 30554 88-22555 . . CAP, SLEEVE 2
29 PAFZZ PAOZZ 6625-00-869-3144 30554 69-599 . . VOLTMETER 1
30 PAFZZ PAOZZ 6625-01-366-0193 30554 88-21103 . . INDICATOR, FUEL LEVEL 1
31 PAFZZ PAOZZ 6685-01-369-6549 30554 88-21105 . . INDICATOR, TEMPERATURE 1

(1) ITEM NO	(2) SMR CODE a. ARMY	(2) SMR CODE b. AIR FORCE	(3) NSN	(4) CAGEC	(5) PART NUMBER	(6) DESCRIPTION AND USABLE ON CODE (UOC)	(7) QTY
32	PAFZZ	PAOZZ	6620-01-368-1531	30554	88-21104	. . INDICATOR, OIL PRESSURE	1
33	PAFZZ	PAOZZ	6645-00-089-8842	30554	73-0507	. . METER, TIME TOTALIZING	1
34	PAFZZ	PAOZZ	6625-00-869-3141	30554	69-574	. . AMMETER	1
35	PAFZZ	PAOZZ	6625-01-366-0192	30554	88-21072	. . METER, ELECTRICAL, FREQ UOC: EMK, 86Q	1
35	PAFZZ	PAOZZ	6625-01-368-7973	30554	88-21073	. . METER, ELECTRICAL, FREQ UOC: YNN, 86R	1
36	PAFZZ	PAOZZ	6625-00-004-8066	30554	69-597	. . AMMETER CURRENT UOC: EMK, 86Q	1
36	PAFZZ	PAOZZ	6625-00-081-5840	30554	69-598	. . AMMETER CURRENT UOC: YNN, 86R	1
37	PAFZZ	PAOZZ	6625-00-003-0975	30554	70-4012	. . WATTMETER	1
38	PAFZZ	PAOZZ	5310-01-307-7914	96906	MS51412-20	. . WASHER, FLAT	3

END OF FIGURE

FIELD AND SUSTAINMENT MAINTENANCE

15 kW 50/60 AND 400 Hz SKID MOUNTED TACTICAL QUIET GENERATOR SETS

GROUP 03 CONTROL BOX ASSEMBLY

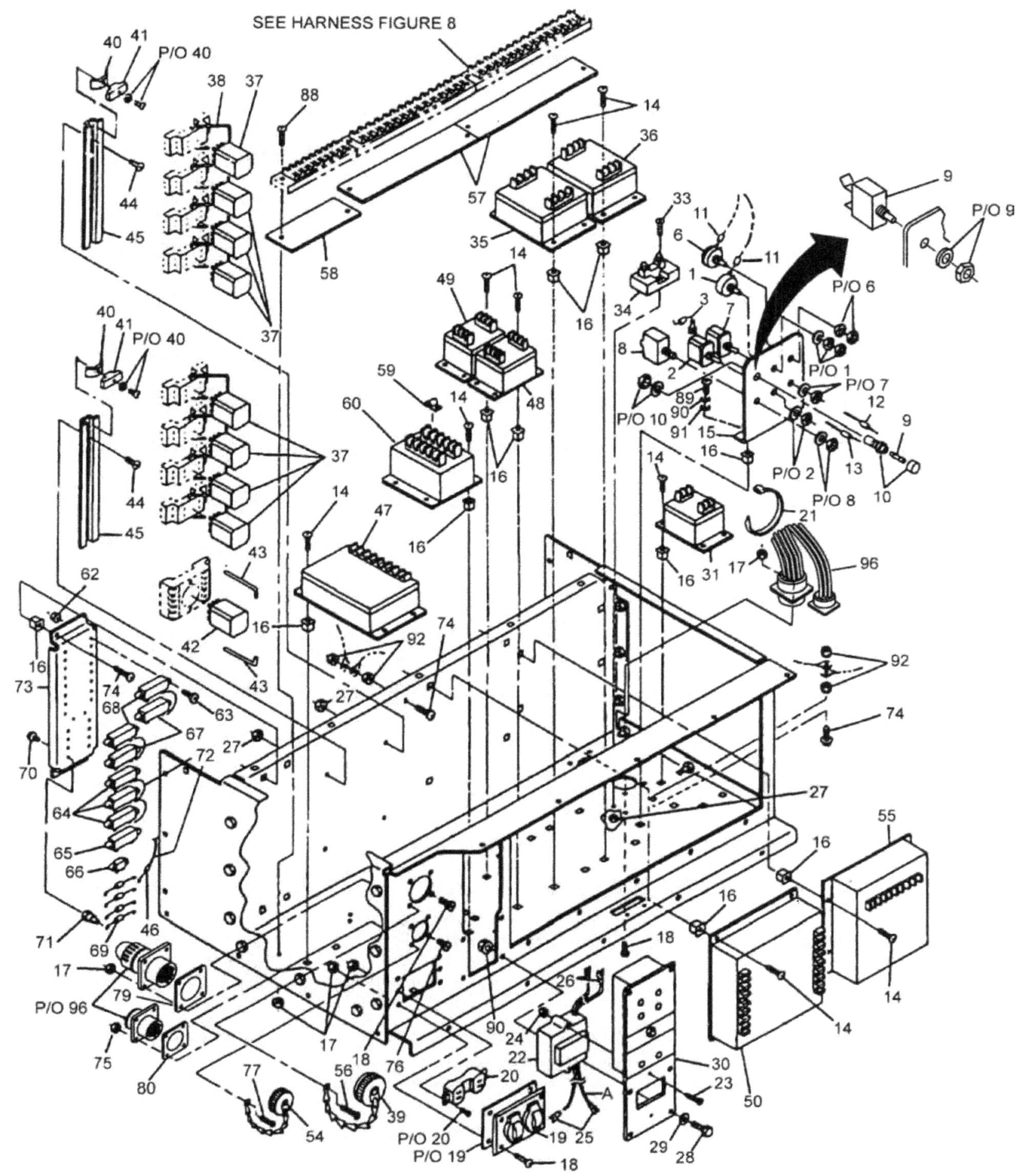

Figure 7. Control Box Assembly (Sheet 1 of 2).

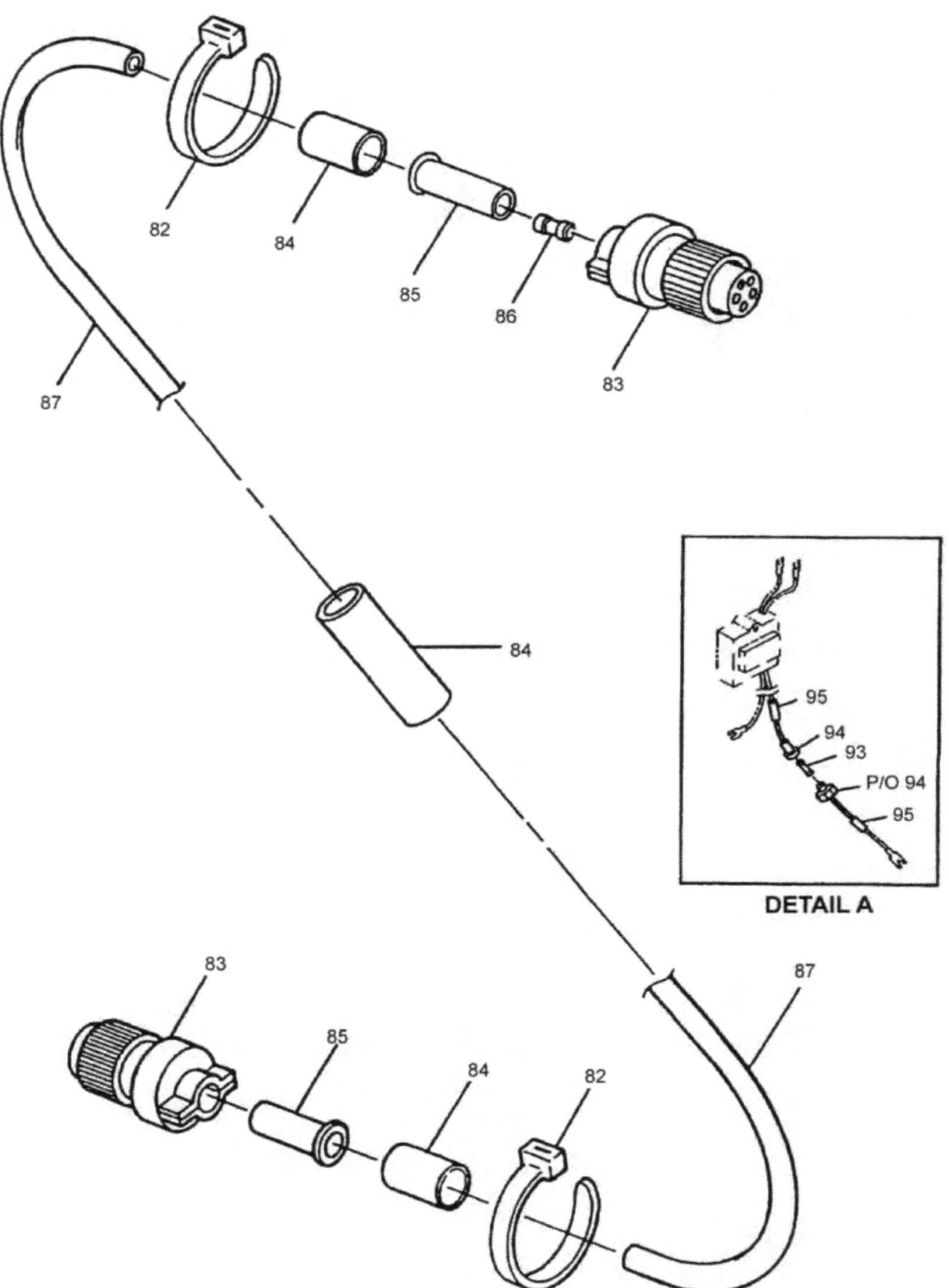

Figure 7. Control Box Assembly (Sheet 2 of 2).

(1) ITEM NO	(2) SMR CODE a. ARMY	b. AIR FORCE	(3) NSN	(4) CAGEC	(5) PART NUMBER	(6) DESCRIPTION AND USABLE ON CODE (UOC)	(7) QTY
						GROUP 03 CONTROL BOX ASSEMBLY	
						FIG. 7 CONTROL BOX ASSEMBLY	
1	PAFZZ	PAOZZ	5905-00-539-2479	30554	88-20592-1	. . RESISTOR, VARIABLE,	1
2	PAFZZ	PAOZZ	5930-01-368-2892	30554	88-22580	. . SWITCH, TOGGLE, FREQ SELECT	1
3	AFFFF	AOO		30554	01-21506-2	. . RESISTOR ASSY UOC: EMK, 86Q	1
3	AFFFF	AOO		30554	01-21506-1	. . RESISTOR ASSY UOC: YNN, 86R	1
4	NOT USED						
5	NOT USED						
6	PAFZZ	PAOZZ	5905-01-336-7533	30554	88-21676-1	. . RESISTOR, VARIABLE,	1
7	PAFZZ	PAOZZ	5930-01-365-9613	30554	88-21080	. . SWITCH, TOGGLE, OVER SPEED RESET	1
8	PAFZA	PAOZA	5925-00-089-3031	30554	88-21155	. . CIRCUIT BREAKER UOC: EMK, YNN	1
9	PAFZZ	PAOZZ	5920-00-539-6920	30554	88-20147-1	. . FUSE, CARTRIDGE UOC: EMK, YNN	1
9	PAFZZ	PAOZZ	5925-00-068-3931	30554	97-24110	. . CIRCUIT BREAKER UOC: 86Q, 86R	1
10	PAFZZ	PAOZZ	5920-00-892-9311	30554	88-20575	. . FUSEHOLDER, EXTRACTOR UOC: EMK, YNN	1
11	MFFZZ	MOO		30554	88-22137-105	. . INSULATION, SLEEVING MAKE FROM P/N M23053/5-105-0 (81349), AS REQUIRED	1
12	MFFZZ	MOO		30554	88-22629-041	. . MARKER, IDENTIFICATION MAKE FROM P/N M23053/5-105-9 (81349), 1.25 INCHES REQUIRED	1
13	MFFZZ	MOO		30554	88-22629-040	. . MARKER, IDENTIFICATION MAKE FROM P/N M23053/5-105-9 (81349), 1.25 INCHES REQUIRED	1
14	PAFZZ	PAOZZ	5305-01-187-5878	30554	69-662-65	. . SCREW, ASSEMBLED	40
15	XDFZZ	XB		30554	88-21593	. . BRACKET, CONTROLS UOC: EMK	1
15	XDFZZ	XB		30554	97-24120	. . BRACKET, CONTROLS UOC: 86Q	1
15	XDFZZ	XB		30554	88-21976	. . BRACKET, CONTROLS UOC: YNN	1
15	XDFZZ	XB		30554	97-24121	. . BRACKET, CONTROLS UOC: 86R	1
16	PAFZZ	PAOZZ	5310-01-366-8134	30554	88-21674-1	. . NUT, CAGE	40
17	PAFZZ	PAOZZ	5310-00-063-7360	30554	69-561-2	. . NUT, PLAIN, ASSEMBLED	24
18	PAFZZ	PAOZZ	5305-00-036-6972	30554	69-662-20	. . SCREW, ASSEMBLED	15
19	PAFZZ	PAOZZ	5975-00-417-0543	30554	70-1580	. . COVER, RECEPTACLE, DUP	1
20	PAFZZ	PAOZZ	5935-01-367-7814	30554	88-20251	. . RECEPTACLE, DUPLEX UOC: EMK, 86Q	1
21	PAFZZ	PAOZZ	5975-00-111-3208	96906	MS3367-5-9	. . STRAP, TIEDOWN, ELECT UOC: EMK, 86Q	10

(1) (2) (3) (4) (5) (6) (7)

SMR CODE

a. b.

ITEM NO ARMY AIR FORCE NSN CAGEC PART NUMBER DESCRIPTION AND USABLE ON CODE (UOC) QTY

22 XBFZZ XBOZZ 5925-01-365-9757 60177 19450-2 . . INTERRUPTER, GROUND, WHEN
REPLACING, USE P/N 88-22825-2
(30554). NO IN-LINE FUSE
REQUIRED
UOC: EMK, 86Q............................ 1

22 PAFZZ PAOZZ 5925-01-493-9106 30554 88-22825-2 . . INTERRUPTER, GROUND
UOC: EMK, 86Q............................ 1

22 PAFFF PAOOO 5925-01-381-5199 60177 29100 . . INTERRUPTER, GROUND, WHEN
REPLACING, USE P/N 88-22826-2
(30554). NO IN-LINE FUSE
REQUIRED
UOC: YNN, 86R............................ 1

22 PAFZZ PAOZZ 5925-01-466-8153 30554 88-22826-2 . . INTERRUPTER, GROUND (NOT
SHOWN)
UOC: YNN, 86R............................ 1

23 PAFZZ PAOZZ 5305-00-150-3408 80205 MS35206-329 ...SCREW, MACHINE
UOC: EMK, 86Q............................ 2

24 PAFZZ PAOZZ 5310-00-081-8087 80205 MS21044N06 . . . NUT, SELF-LOCKING, HEX .... 1 25 PAFZZ PAOZZ 5940-01-259-2190 30554 88-20274-11 ...TERMINAL, LUG #8,22-18 AWG
UOC: EMK, 86Q............................ 2

26 PAFZZ PAOZZ 5940-01-110-6423 30554 88-20274-5 ...TERMINAL, LUG #6,16-14 AWG
UOC: EMK, 86Q............................ 2

27 PAFZZ PAOZZ 5310-00-052-3632 30554 69-561-3 . . NUT, PLAIN, ASSEMBLED 6

28 PAFZZ PAOZZ 5306-01-156-7663 30554 88-20260-21 . . SCREW, HEX WASHERHEAD 6

29 PAFZZ PAOZZ 5310-01-234-9416 96906 MS51412-2 . . WASHER, FLAT 6

30 PBFZZ PBOZZ 6625-01-381-7445 30554 88-21142 . . INDICATOR, FAULT, LOC .... 1 31 PAFZZ PAOZZ 6695-01-368-7114 30554 88-21134 . . TRANSDUCER, FREQUENCY
UOC: EMK, 86Q............................ 1

31 PAFZZ PAOZZ 6695-01-369-9312 30554 88-21135 . . TRANSDUCER, FREQUENCY
UOC: YNN, 86R............................ 1

32 NOT USED

33 PAFZZ PAOZZ 5305-00-038-3145 30554 69-662-37 . . SCREW, ASSEMBLED 2

34 PBFZZ PBOZZ 6625-01-381-8195 30554 88-21165 . . SHUNT, INSTRUMENT 1

35 PAFZZ PAOZZ 5945-01-366-2725 30554 88-21138 . . RELAY, OVER/UNDER VOL 1

36 PAFZZ PAOZZ 5945-01-366-2726 30554 88-21141 . . RELAY, ELECTROMAGNET 1

37 PAFZZ PAOZZ 5945-00-458-3351 30554 88-21085 . . RELAY, ELECTROMAGNET 8

38 XDFZZ XB 5342-01-078-9038 30554 88-22768 . . CLIP, SPRING TENSION 8

39 PAFZZ PAOZZ 5935-01-114-5354 30554 88-22672 . . COVER, ELECTRICAL CO 4

40 PAFZA PAOZA 5340-01-078-9038 30554 88-22768 . . MOUNTING CLIP 4

41 PAFZZ PAOZZ 5970-01-280-0362 30554 88-22767 . . INSULATOR .... 4 42 PAFZZ PAOZZ 5945-01-365-9954 30554 88-21896 . . RELAY ,ELECTROMAGNET
UOC: EMK, 86Q............................ 1

42 PAFZZ PAOZZ 5945-01-369-0791 30554 88-22463 . . RELAY, ELECTROMAGNET
UOC: YNN, 86R............................ 1

43 XDFZZ XB 30554 88-21897-2 . . CLIP, RELAY SOCKET 2

44 PAFZZ PAOZZ 5305-00-038-3103 30554 69-662-35 . . SCREW, ASSEMBLED 4

45 XDFZZ XB 5340-01-470-2973 30554 88-20258 . . TRACK, SOCKET MOUNT .... 2 46 MFFZZ MOO 30554 88-22629-021 . . MARKER, IDENTIFICATION MAKE
FROM P/N M23053/5-105-9 (81349),
1.25 INCHES REQUIRED 1

47 PAFZZ PAOZZ 6695-01-367-9722 30554 88-21133 . . TRANSDUCER, WATT 1

48 PAFZZ PAOZZ 5945-01-366-2727 30554 88-21145 . . RELAY, PERMISSIVE PARAL 1

49 PAFZZ PAOZZ 5945-01-366-2728 30554 88-21140 . . RELAY, REVERSE POWER .... 1 50 PBFZZ PBOZZ 2910-01-368-7644 30554 88-21860 . . GOVERNOR, DIESEL ENG
UOC: EMK 1

(1)	(2) SMR CODE		(3)	(4)	(5)	(6)	(7)
	a.	b.					
ITEM NO	ARMY	AIR FORCE	NSN	CAGEC	PART NUMBER	DESCRIPTION AND USABLE ON CODE (UOC)	QTY

50 PBFZZ PBOZZ 30554 97-24134 . . GOVERNOR, DIESEL ENG
UOC: 86Q.................................... 1
50 PBFZZ PBOZZ 2910-01-371-4356 30554 88-21861 . . GOVERNOR, DIESEL ENG
UOC: YNN, 86R............................ 1
51 NOT USED
52 NOT USED
53 NOT USED
54 PAFZZ PAOZZ 5935-01-175-8419 96906 MS25043-18DA . . COVER, ELECTRICAL CO 1
55 PAFZZ PAOZZ 6110-01-363-0493 30554 01-21501-1 . . REGULATOR, VOLTAGE
UOC: EMK, 86Q............................ 1
55 PAFZZ PAOZZ 6110-01-379-7187 30554 01-21501-2 . . REGULATOR, VOLTAGE
UOC: EMK, 86Q............................ 1
55 PAFZZ PAOZZ 6110-01-368-7123 30554 01-21507-1 . . VOLTAGE REGULATOR
UOC: YNN, 86R............................ 1
56 PAFZZ PAOZZ 5305-00-218-4864 30554 69-662-22 . . SCREW, ASSEMBLED 7
57 XDFZZ XB 5940-01-531-5660 30554 88-21182-22 . . MARKER, STRIP 2
58 XDFZZ XB 30554 88-21182-12 . . MARKER, STRIP 1
59 PAFZZ PAOZZ 5999-01-366-2621 30554 88-21183-2 . . CONTACT, ELECTRICAL 2
60 PBFZZ PBOZZ 6625-01-367-8436 30554 88-21144 . . LOAD MEASUREMENT UNIT 1
61 AFFFF AOO 30554 88-22106 . . RESISTOR, DIODE ASSY (NOT SHOWN) 1
62 PAFZZ PAOZZ 5310-00-836-3520 30554 69-561-1 . . . NUT, PLAIN, ASSEMBLED 18
63 PAFZZ PAOZZ 5305-00-224-1092 30554 69-662-5 ...SCREW, ASSEMBLED 18
64 PAFZZ PAOZZ 5905-01-368-2538 30554 88-20262 ...RESISTOR,FIXED,WIRE 3
65 PAFZZ PAOZZ 5905-01-365-6585 30554 88-20263 ...RESISTOR,FIXED,WIRE 1
66 PAFZZ PAOZZ 5905-01-368-2539 30554 88-20265 ...RESISTOR,FIXED,WIRE 1
67 PAFZZ PAOZZ 5905-01-366-7074 30554 88-20266 ...RESISTOR,FIXED,WIRE 2
68 PAFZZ PAOZZ 5905-01-368-2540 30554 88-20267 ...RESISTOR,FIXED,WIRE 2
69 PAFZZ PAOZZ 5961-00-295-5757 50434 1901-0759 . . . SEMICONDUCTOR DEVICE 4
70 PAFZZ PAOZZ 5305-01-365-9390 30554 88-20299 ...SCREW, ASSEMBLED 8
71 PAFZZ PAOZZ 5940-00-954-3558 30554 88-20272 ...TERMINAL, INSULATED 8
72 MFFZZ MOO 30554 88-22106-11 . . . WIRE, ELECTRIC MAKE FROM P/N M5086/2-20-9 (81349), AS REQUIRED 1
73 XDFZZ XB 30554 88-21971 ...BRACKET,BURDEN RES 1
74 PAFZZ PAOZZ 5305-01-187-5878 30554 69-662-65 . . SCREW, ASSEMBLED 2
75 PAFZZ PAOZZ 5310-00-836-3520 30554 69-561-1 . . NUT, PLAIN, ASSEMBLED 4
76 PAFZZ PAOZZ 5305-00-224-1092 30554 69-662-5 . . SCREW, ASSEMBLED 3
77 PAFZZ PAOZZ 5305-01-204-4683 30554 69-662-7 . . SCREW, ASSEMBLED 1
78 NOT USED
79 PAFZZ PAOZZ 5330-00-914-7651 96906 MS51007-12 . . GASKET
UOC: EMK, 86Q............................ 1
80 PAFZZ PAOZZ 5330-00-079-7840 96906 MS51007 . . GASKET
UOC: EMK, 86Q............................ 1
81 MFFFF MFOFF 6150-01-406-9533 30554 88-22209 . . CABLE ASSEMBLY (NOT SHOWN) 1
82 PAFZZ PAOZZ 5975-00-074-2072 96906 MS3367-1-9 ...STRAP, TIEDOWN,ELECT 2
83 PAFZA PAOZA 5935-00-233-3990 96906 MS3106F18-1S ...CONNECTOR,PLUG,ELEC 2
84 MFFZZ MOO 30553 88-22209-7 ...INSULATION, SLEEVING MAKE FROM P/N M23053/5-108-9 (81349), AS REQUIRED 1
85 PAFZZ PAOZZ 5365-00-663-2125 96906 MS3420-10 ...BUSHING,NONMETALLIC 2
86 PAFZZ PAOZZ 5935-00-565-9503 96906 MS25251-12 ...PLUG,END SEAL, ELECT 1

(1)	(2) SMR CODE		(3)	(4)	(5)	(6)	(7)
ITEM NO	a. ARMY	b. AIR FORCE	NSN	CAGEC	PART NUMBER	DESCRIPTION AND USABLE ON CODE (UOC)	QTY

87 MFFZZ MOO 30554 88-21995 ...CABLE, MULTICONDUCTOR
MAKE FROM P/N 8407 (16498), 300 INCHES REQUIRED 1

88 PAFZZ PAOZZ 5305-00-218-4864 30554 69-662-22 . . SCREW, ASSEMBLED 6

89 PAFZZ PAOZZ 5306-01-156-7663 30554 88-20260-21 . . SCREW, HEX WASHERHEAD 2

90 PAFZZ PAOZZ 5310-00-274-8710 80205 MS35338-62 . . WASHER, LOCK 2

91 PAFZZ PAOZZ 96906 MS51512-2 . . WASHER, FLAT 2

92 PAFZZ PAOZZ 5310-01-012-3595 30554 69-561-6 . . NUT, PLAIN, ASSEMBLED 2

93 PAFZZ PAOZZ 5920-00-243-3787 81349 F03A250V10AS . . FUSE, CARTRIDGE 1

94 PAFZZ PAOZZ 5920-01-396-1989 75915 150145 . . FUSEHOLDER, EXTRACTOR 1

95 PAFZZ PAOZZ 5940-00-478-0037 56501 RBB25 . . SPLICE, CONDUCTOR .... 1 96 PBFFF PBOOO 6150-01-383-6511 30554 88-22303 . . WIRE HARNESS, BRANCH (SEE
FIGURE 8 FOR PARTS BREAK-DOWN)
UOC: EMK .... REF

96 PBFFF PBOOO 6150-01-384-0013 30554 88-22664 . . WIRING HARNESS, BRANCH (SEE
FIGURE 8 FOR PARTS BREAK-DOWN)
UOC: YNN .... REF

96 PBFFF PBOOO 30554 97-24049 . . WIRING HARNESS, BRANCH (SEE
FIGURE 8 FOR PARTS BREAK-DOWN)
UOC: 86Q.................................. REF

96 PBFFF PBOOO 30554 97-24060 . . WIRING HARNESS, BRANCH REF
(SEE FIGURE 8 FOR PARTS BREAK-DOWN)
UOC: 86R ....

END OF FIGURE

FIELD AND SUSTAINMENT MAINTENANCE

15 kW 50/60 AND 400 Hz SKID MOUNTED TACTICAL QUIET GENERATOR SETS
GROUP 03 CONTROL BOX ASSEMBLY: CONTROL BOX HARNESS ASSEMBLY

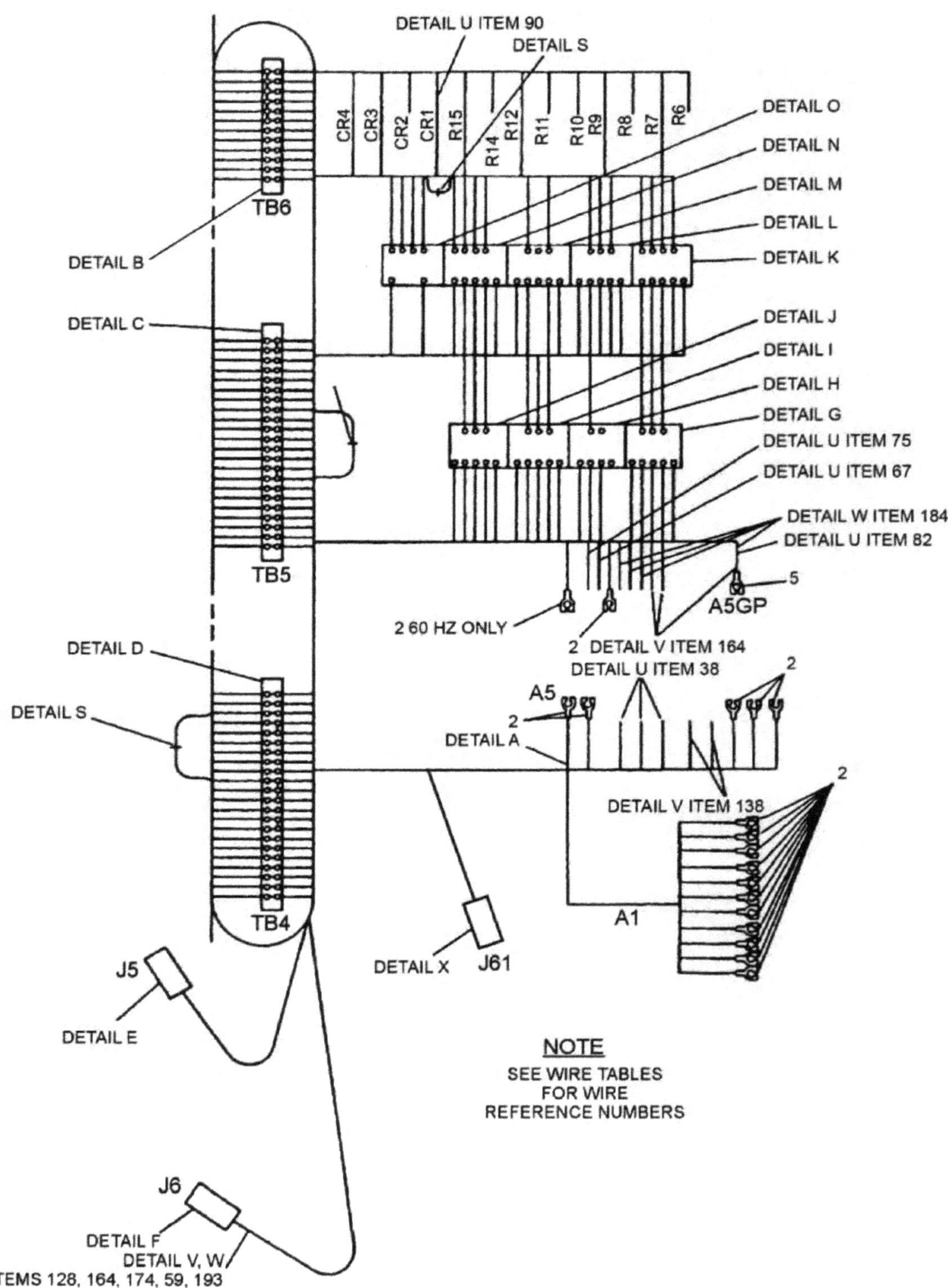

Figure 8. Control Box Harness Assembly (Sheet 1 of 25).

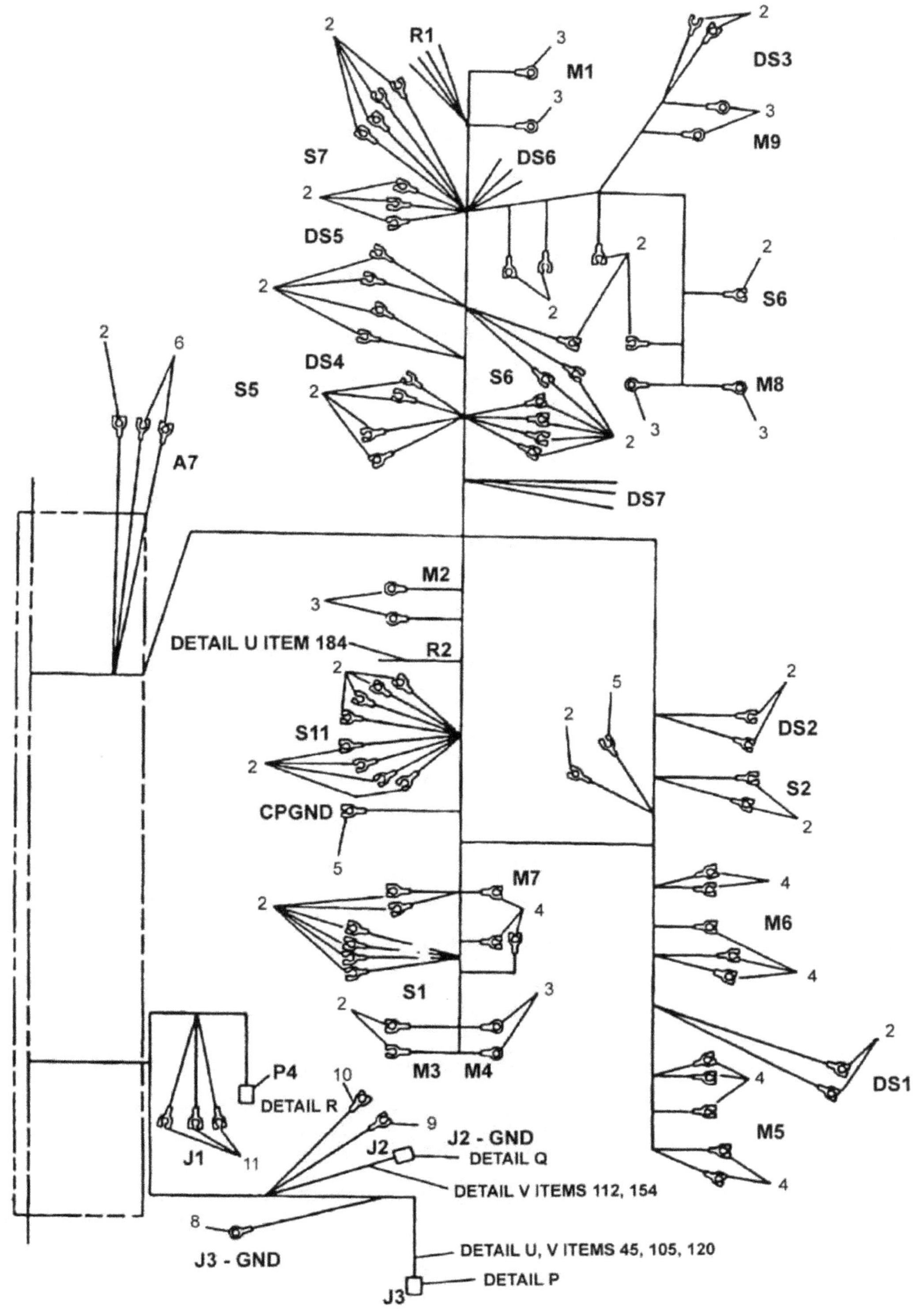

Figure 8. Control Box Harness Assembly (Sheet 2 of 25).

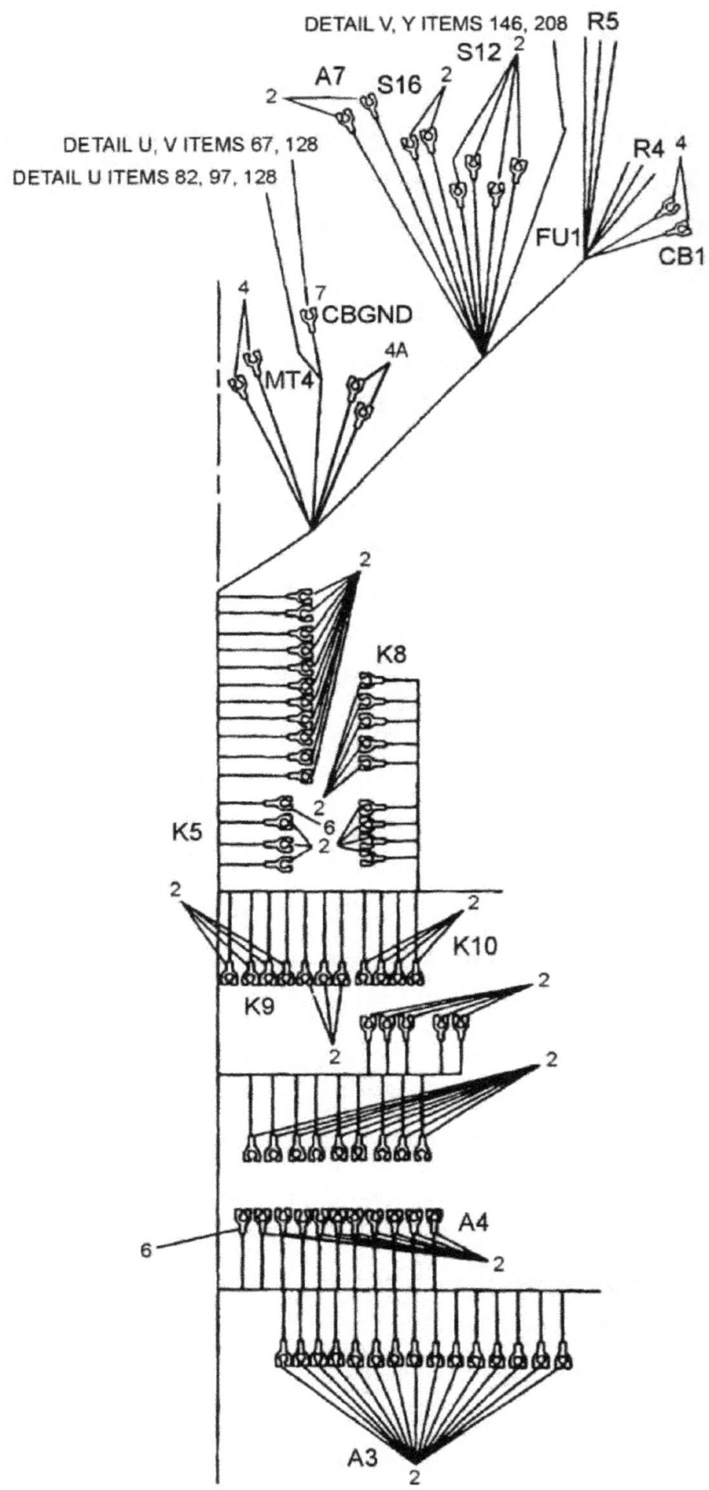

Figure 8. Control Box Harness Assembly (Sheet 3 of 25).

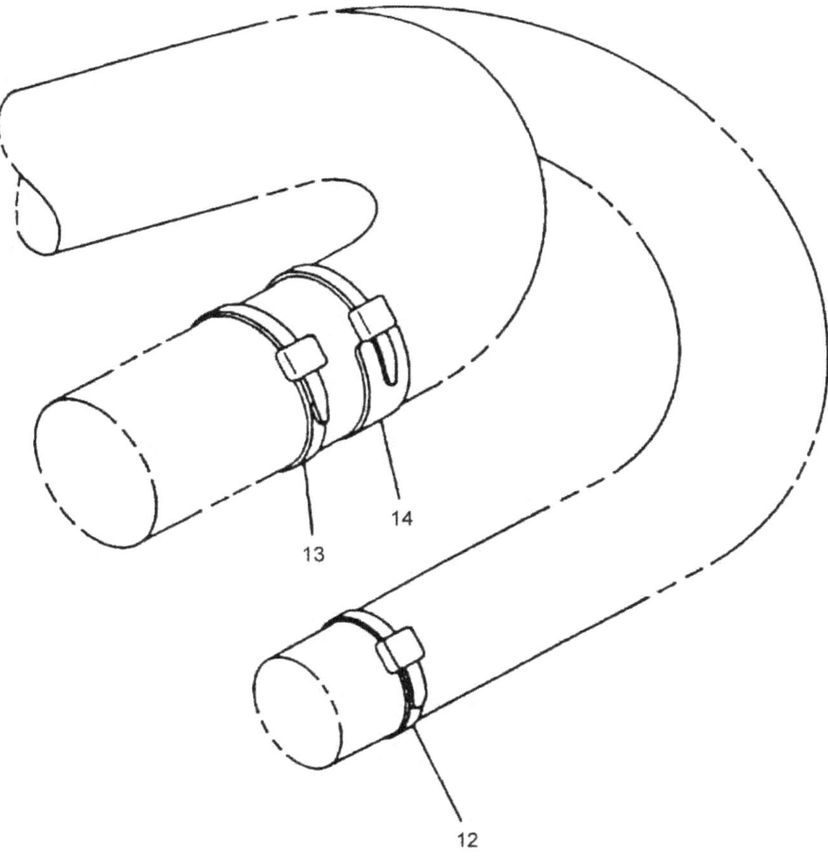

DETAIL A
(TYPICAL INSTALLATION)

Figure 8. Control Box Harness Assembly (Sheet 4 of 25).

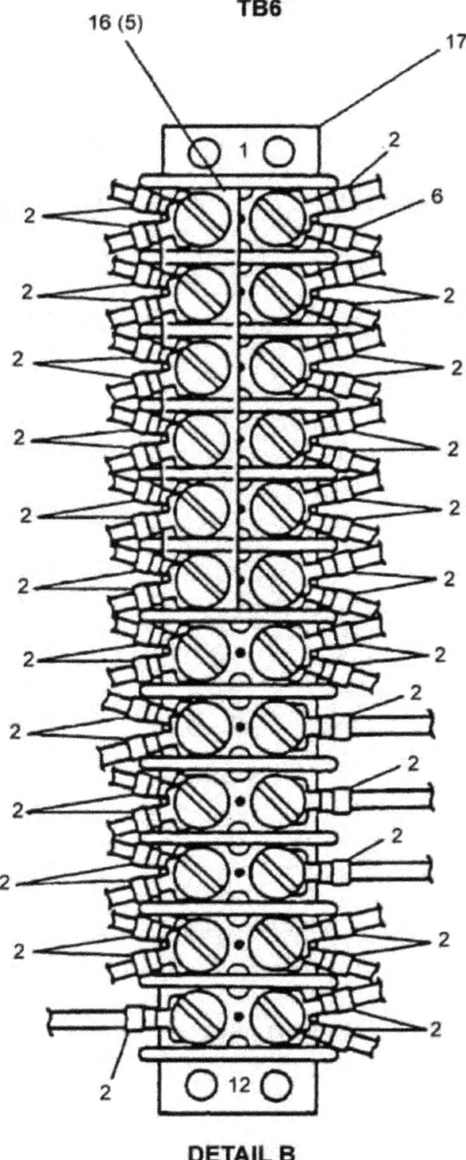

Figure 8. Control Box Harness Assembly (Sheet 5 of 25).

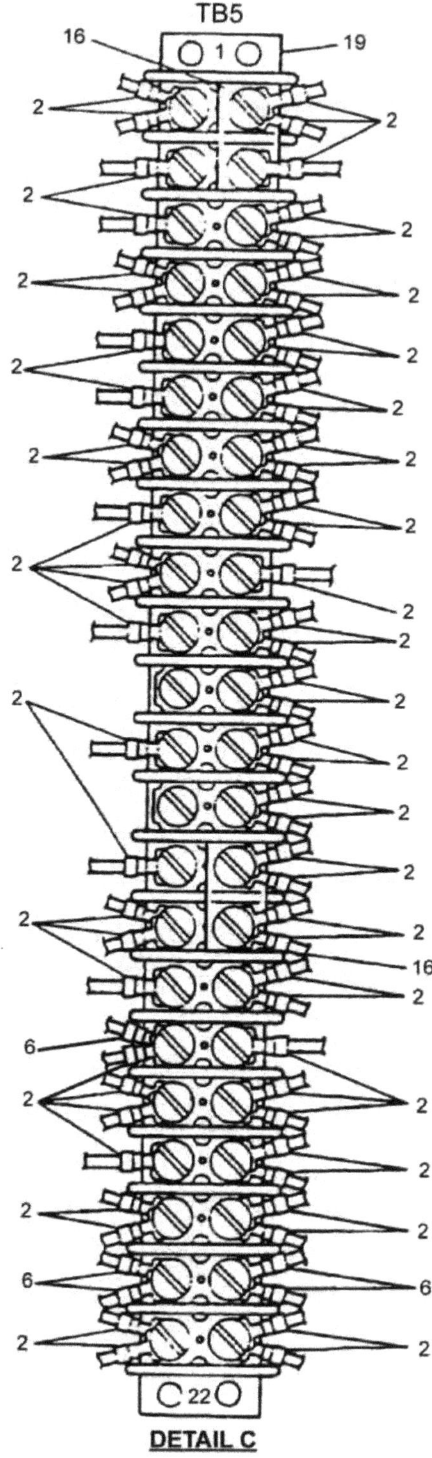

Figure 8. Control Box Harness Assembly (Sheet 6 of 25).

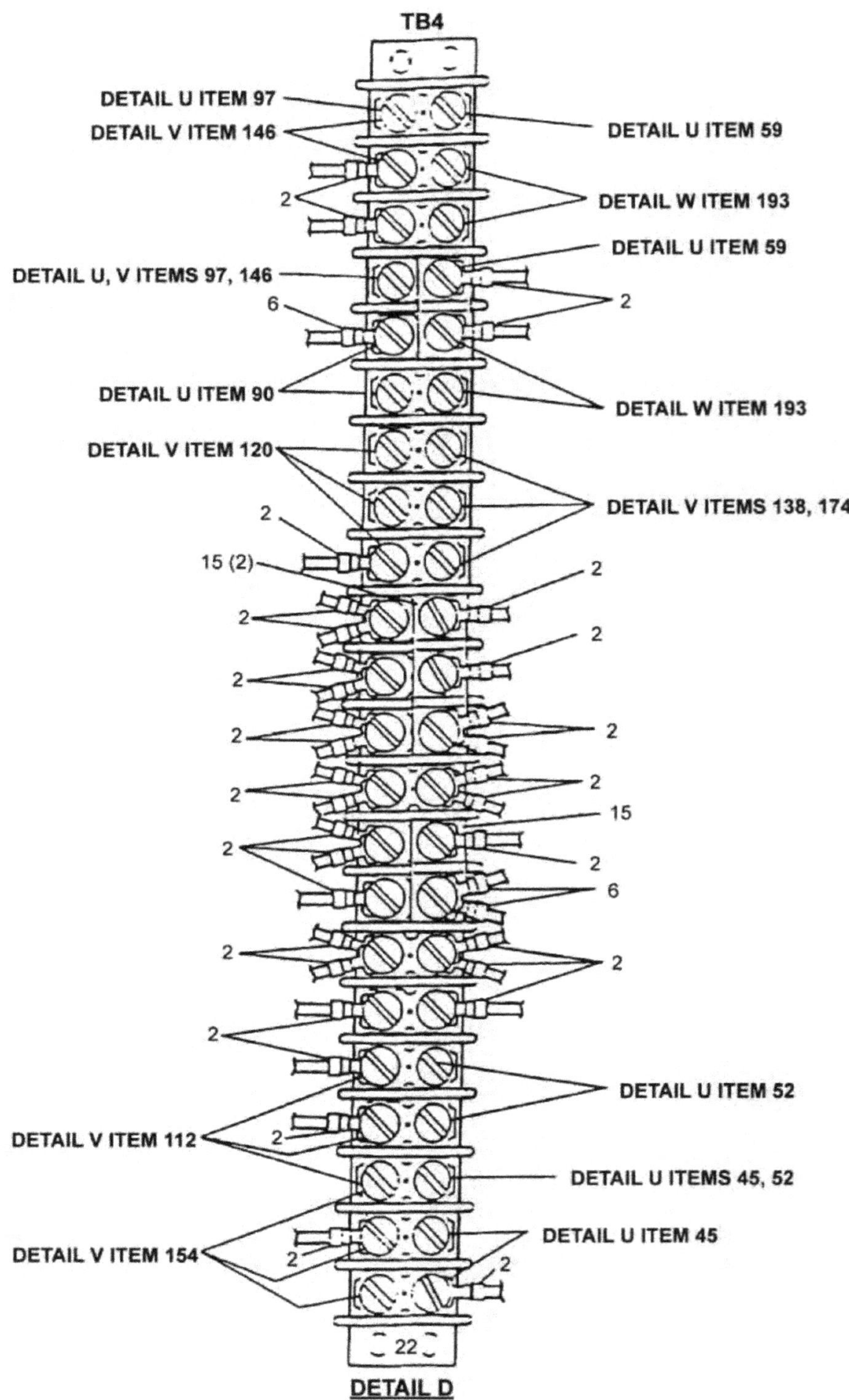

Figure 8. Control Box Harness Assembly (Sheet 7 of 25).

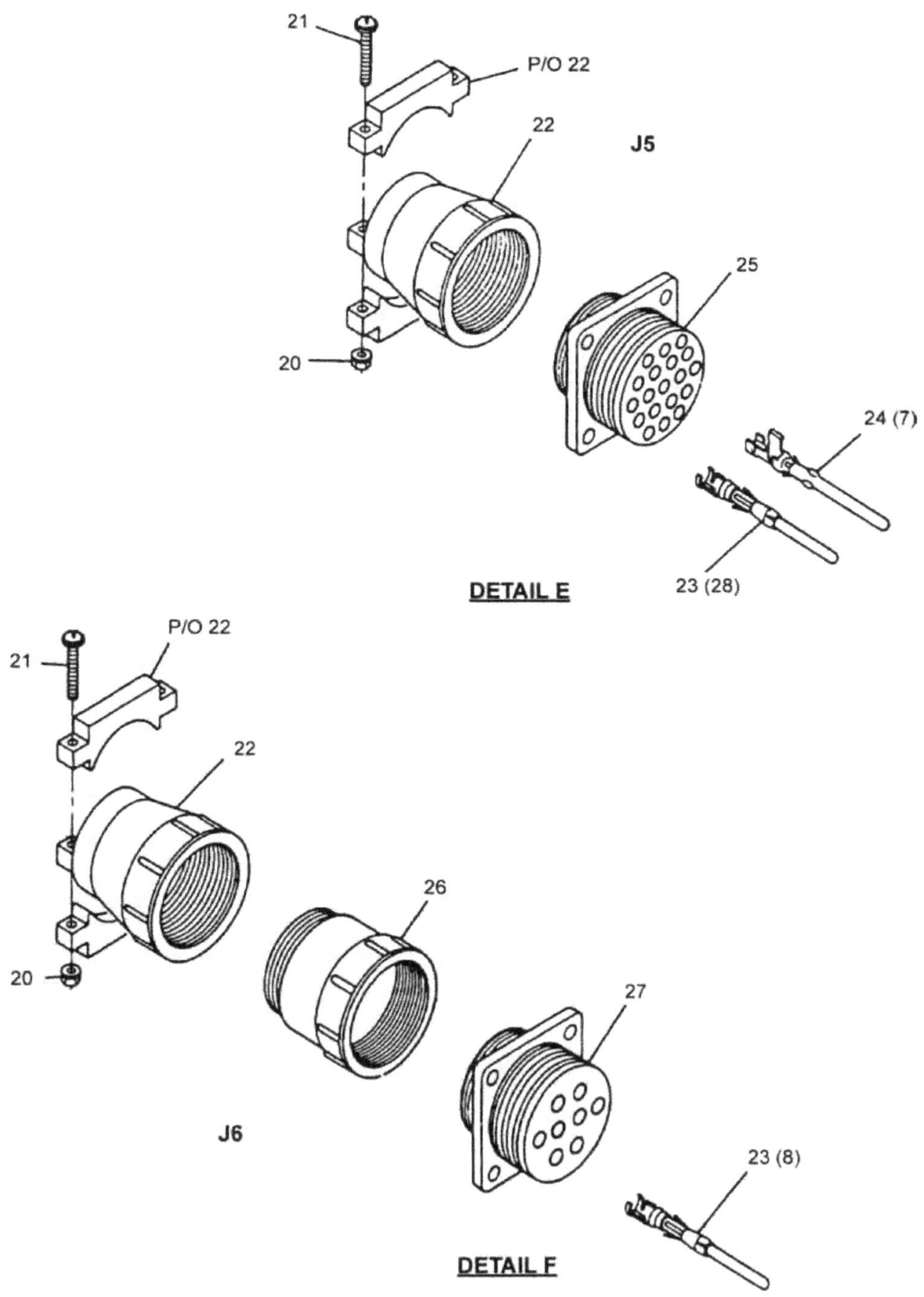

Figure 8. Control Box Harness Assembly (Sheet 8 of 25).

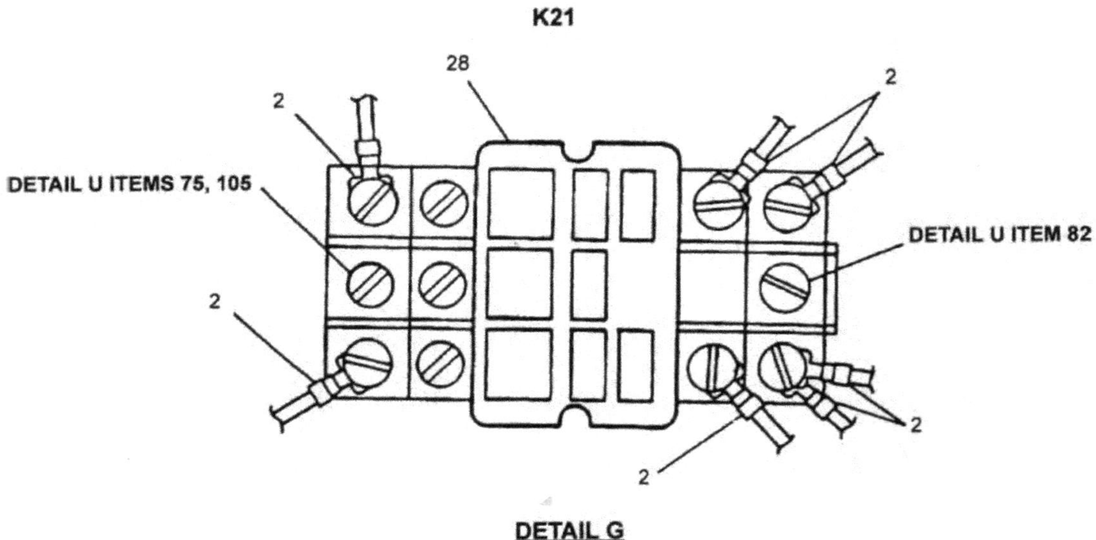

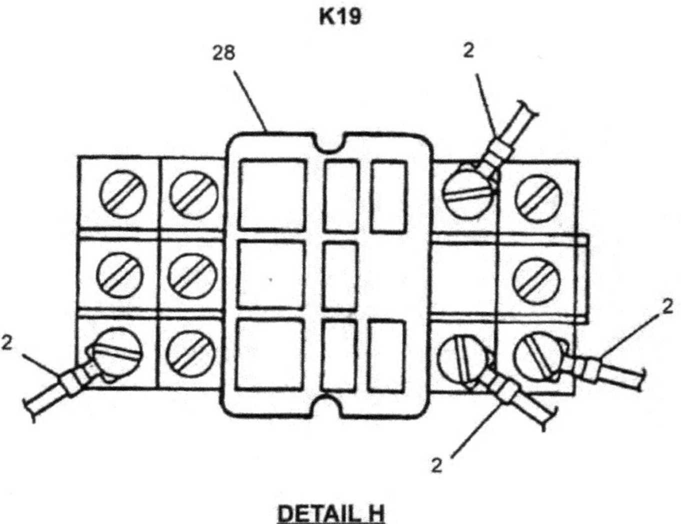

Figure 8. Control Box Harness Assembly (Sheet 9 of 25).

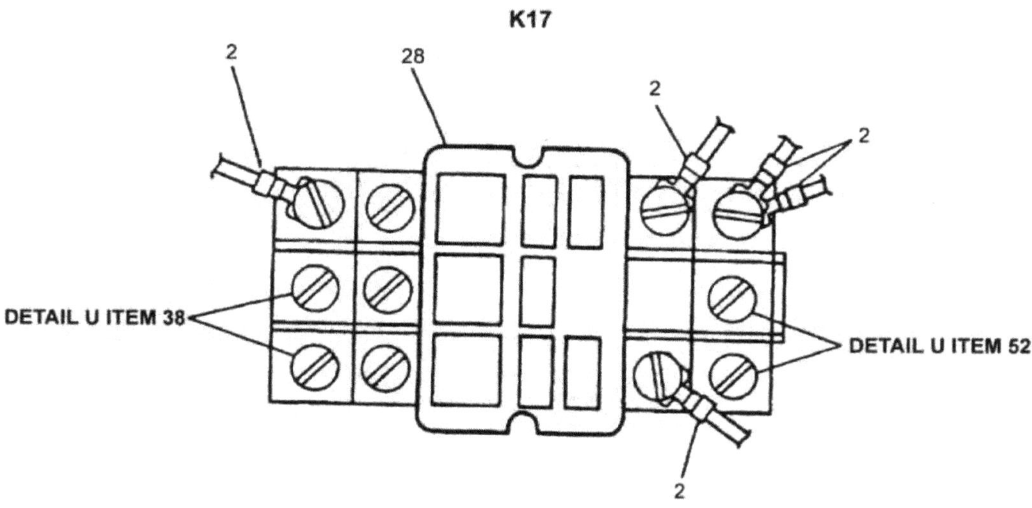

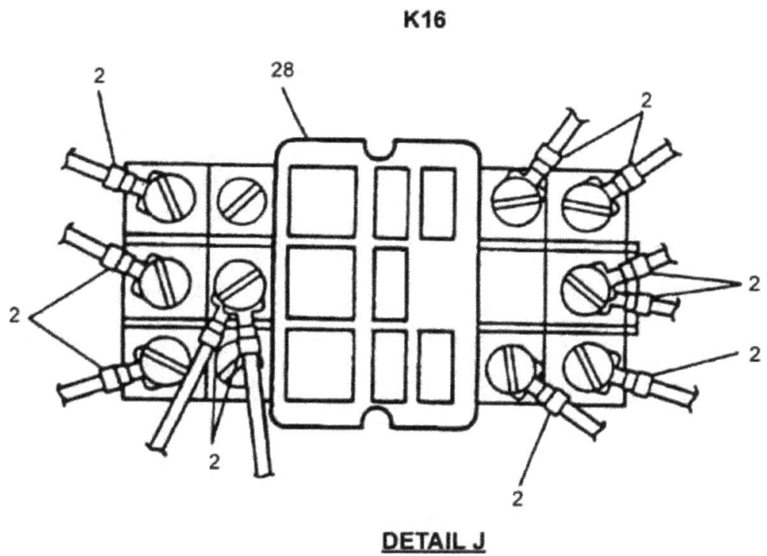

Figure 8. Control Box Harness Assembly (Sheet 10 of 25).

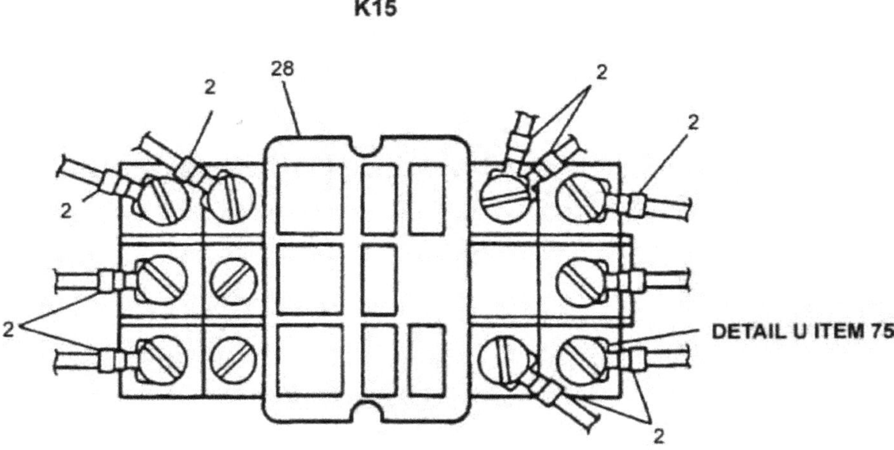

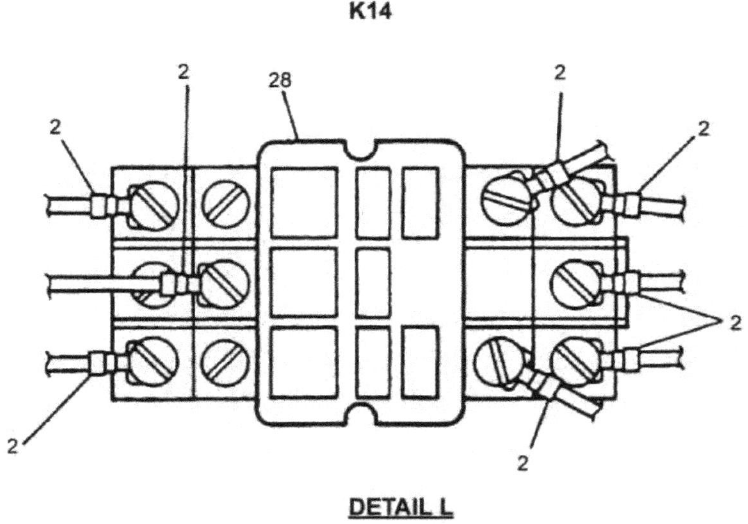

Figure 8. Control Box Harness Assembly (Sheet 11 of 25).

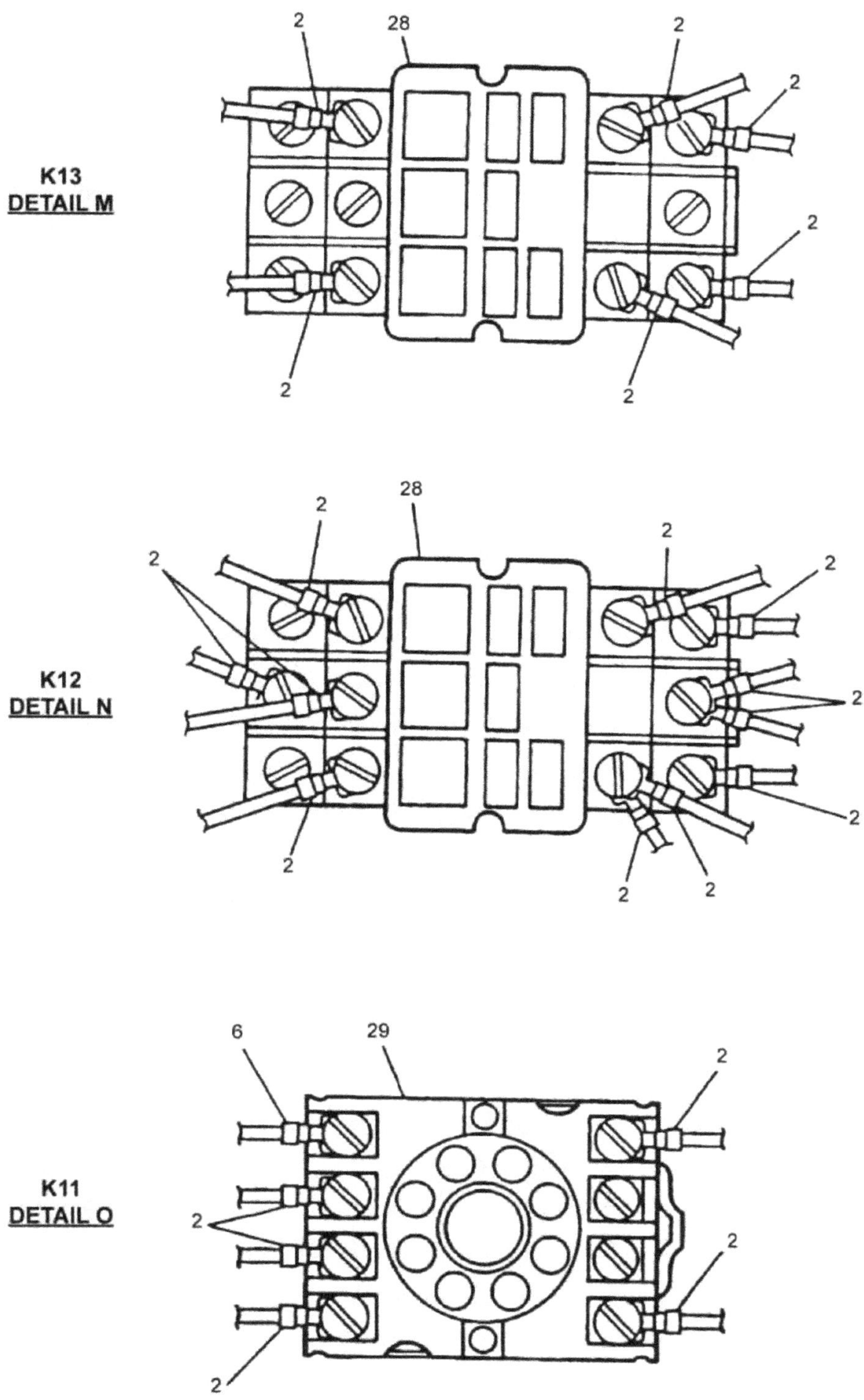

Figure 8. Control Box Harness Assembly (Sheet 12 of 25).

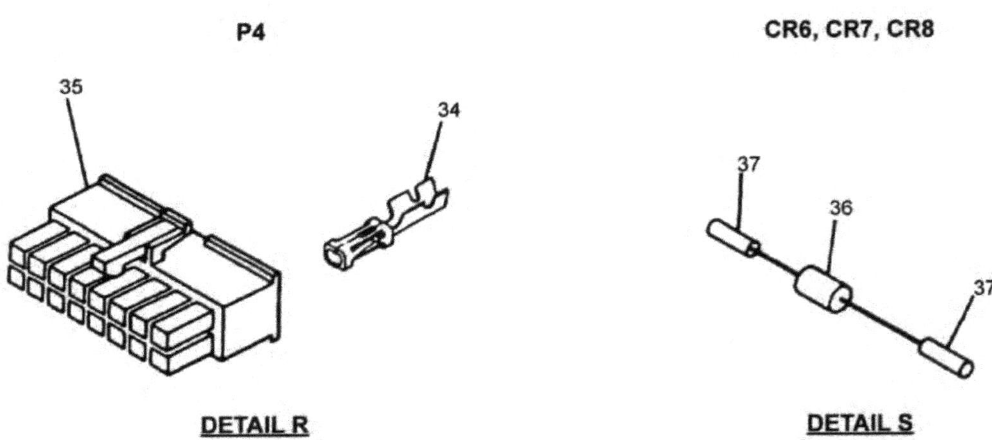

Figure 8. Control Box Harness Assembly (Sheet 13 of 25).

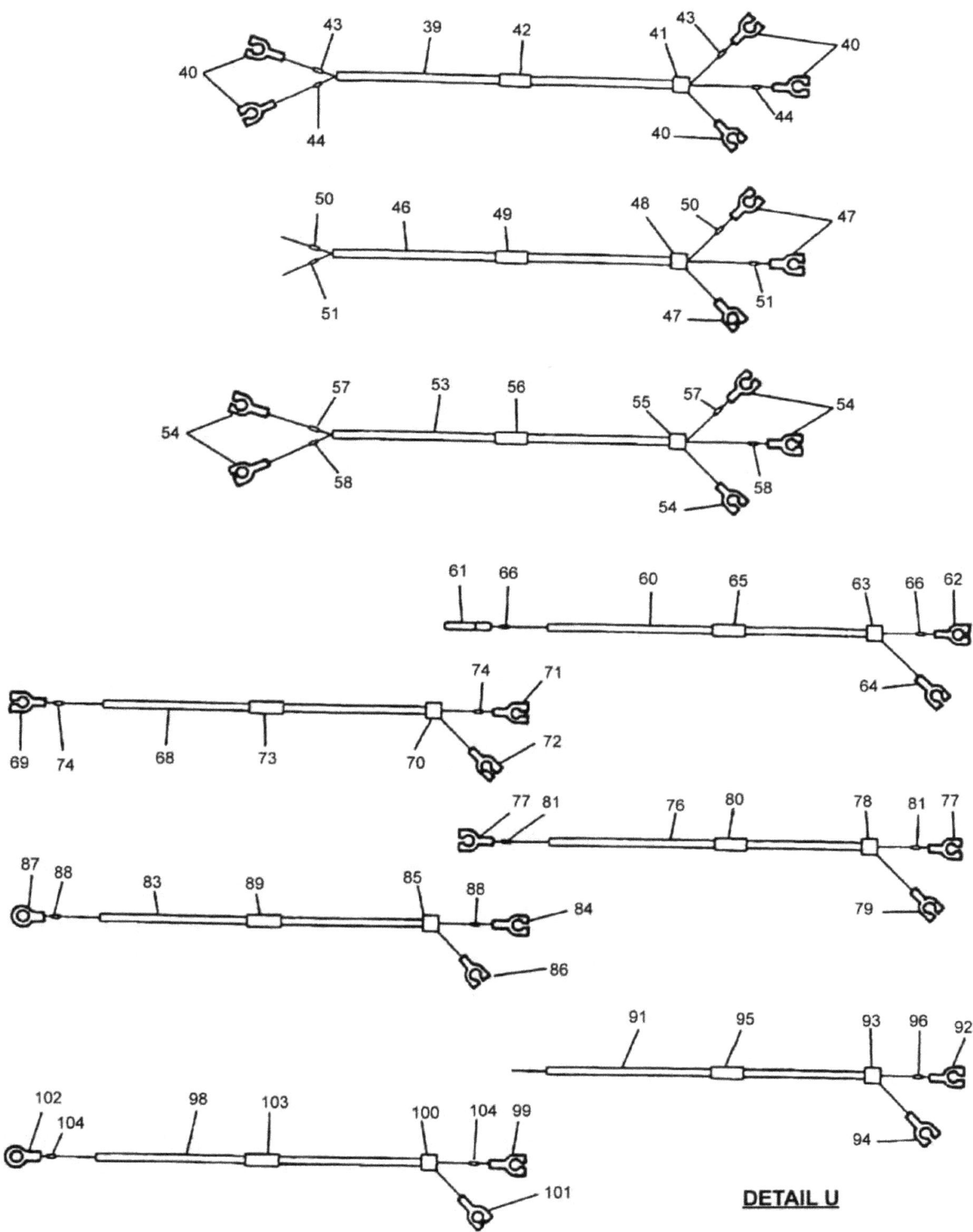

Figure 8. Control Box Harness Assembly (Sheet 14 of 25).

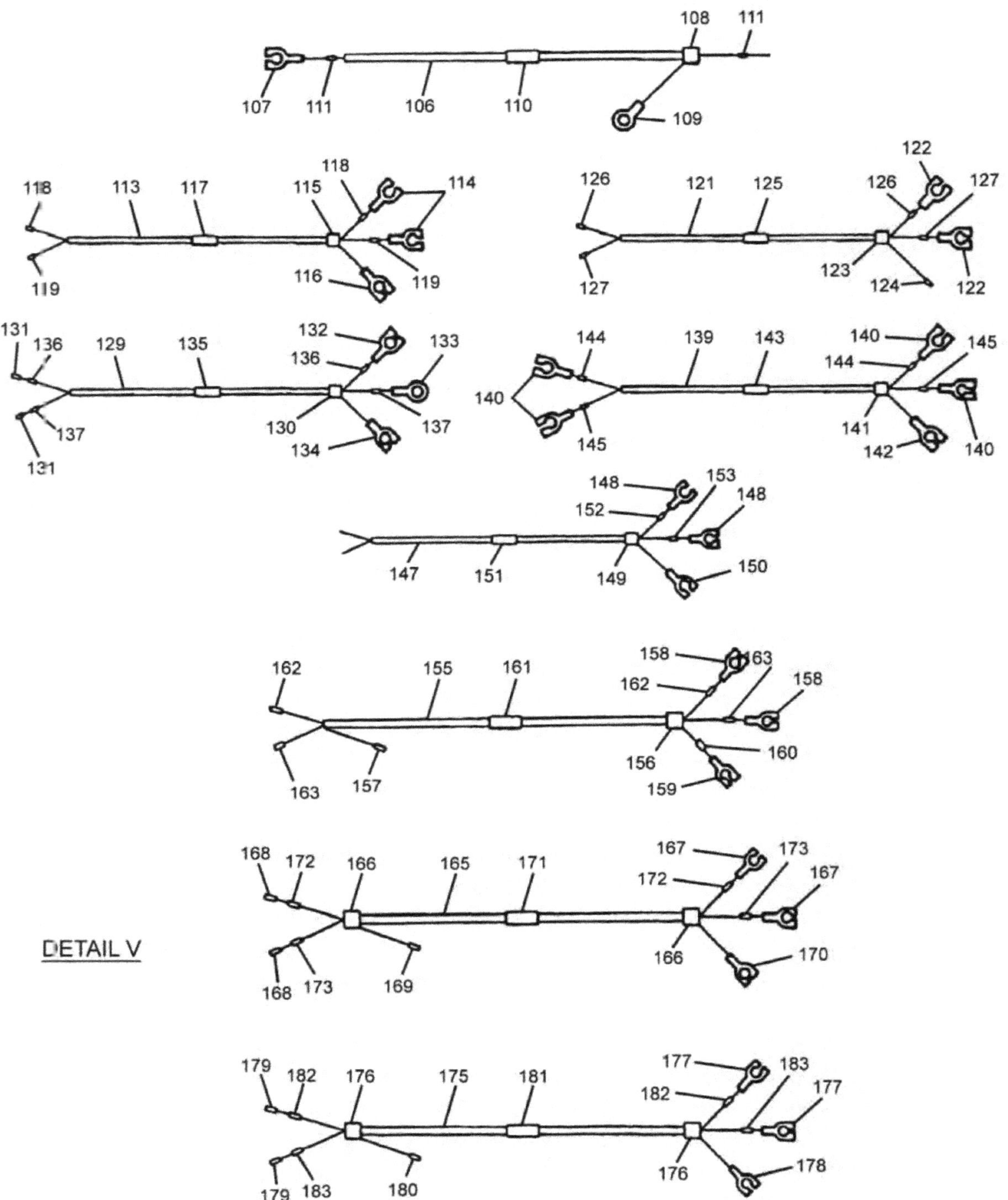

Figure 8. Control Box Harness Assembly (Sheet 15 of 25).

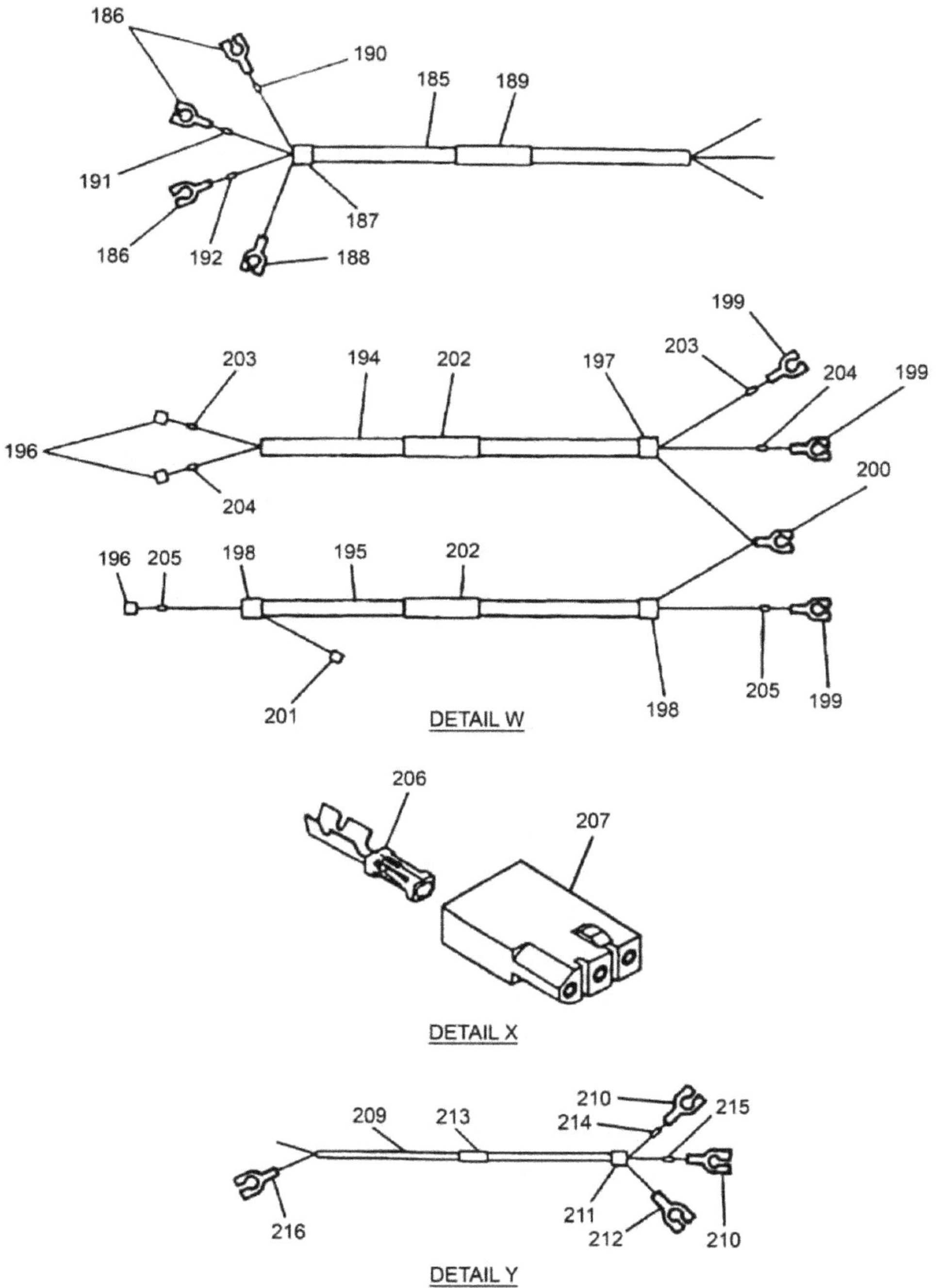

Figure 8. Control Box Harness Assembly (Sheet 16 of 25).

WIRE REF NO.	WIRE MARKING NO.	FROM	FIG.8 ITEM NO.	TO	FIG.8 ITEM NO.	FIG.8 ITEM NO.	MARKING COLOR	WIRE LENGTH
1	100AA20	J3-GND	8	TB6-1	2	15	RED	36.5
2	100AB20	M3 (–)	2	TB6-2	2	15	RED	37.0
3	100AC20	K12-B	2	TB6-3	2	15	RED	14.5
4	100AD20	J2-GND	9	TB6-6	2	15	RED	37.0
5	100AE20	K13-B	2	TB6-4	2	15	RED	22.0
6	100AF20	P4-5	34	TB6-5	2	15	RED	20.0
7	100AG20	P4-12	34	TB6-5	2	15	RED	20.0
8	100AH20	DS1-2	2	TB6-1	2	15	RED	40.0
9	100AJ20	DS2-2	2	TB6-2	2	15	RED	35.0
10	100AK20	DS3-2	2	TB6-3	2	15	RED	43.0
11	100AL20	M6-G	4	M5-G	4	15	RED	9.0
12	100AM20	M7-G	4	M6-G	4	15	RED	17.0
13	100AN20	M5-G	4	TB6-1	2	15	RED	41.0
14	100AP20	DS6-3	TIN	TB6-6	2	15	RED	30.0
15	100AR12	J2-E	–	J2-GND	10	33	RED	6.0
16	100AS20	TB6-2	2	K9-4	2	15	RED	24.0
17	100AT20	K9-4	2	K10-4	2	15	RED	5.0
18	100AV16	J1-1	11	TB6-1	6	18	RED	21.5
19	100AY20	DS7-3	TIN	TB6-3	2	15	RED	40.0
20	100AZ20	S7-8	2	TB6-2	2	15	RED	43.0
21	100F20	TB6-6	2	TB4-5	2	15	RED	19.0
22	100G16	TB4-5	6	CBGND	7	18	RED	14.0
23	100H20	A5-GP	5	TB4-4	2	15	RED	17.0
24	100T20	K10-7	2	TB6-5	2	15	RED	25.0
25	100U20	K14-7	2	TB6-5	2	15	RED	17.0
26	100V20	K21-B	2	TB6-4	2	15	RED	24.0
27	100W20	K15-9	2	TB6-4	2	15	RED	26.0
28	100X20	CPGND	4	TB6-3	2	15	RED	33.0
29	100Y20	K16-B	2	TB6-4	2	15	RED	18.5
30	100Z20	TB6-6	2	TB4-9	2	15	RED	21.5
31	101C20	J5-1	23	TB4-17	2	15	BLACK	13.0
32	101D20	S6-10	2	TB4-17	2	15	BLACK	32.5
33	100BC20	J3-B	–	K10-4	2	15	BLACK	43.0
34	102C20	J5-2	23	TB6-11	2	15	BLACK	31.0
35	102D20	K10-1	2	TB6-11	2	15	BLACK	31.5
36	102E20	S6-6	2	TB6-11	2	15	BLACK	34.5
37	102F20	DS5-1	2	TB6-11	2	15	BLACK	31.0
38	NOT USED							
39	103C20	J5-3	23	TB6-10	2	15	BLACK	31.0
40	103D20	DS4-1	2	TB6-10	2	15	BLACK	29.0
41	103E20	S6-2	2	TB6-10	2	15	BLACK	35.0
42	NOT USED							
43	185K20	J5-4	23	TB5-20	2	15	RED	24.5
44	NOT USED							
45	185J20	K17-A	2	TB5-20	2	15	RED	10.0
46	185H20	J3-V	–	TB5-20	2	15	RED	38.0
47	105B20	J5-5	23	TB5-3	2	15	RED	34.0

Figure 8. Control Box Harness Assembly (Sheet 17 of 25).

WIRE REF NO.	WIRE MARKING NO.	FROM	FIG.8 ITEM NO.	TO	FIG.8 ITEM NO.	FIG.8 ITEM NO.	MARKING COLOR	WIRE LENGTH
48	NOT USED							
49	105D20	K13-1	2	TB5-3	2	15	RED	12.0
50	105E20	DS7-2	TIN	TB5-3	2	15	RED	46.0
51	106B16	J5-6	24	K11-7	6	18	BLACK	34.0
52	107D16	J5-36	24	TB5-21	6	18	BLACK	23.0
53	107E16	A4-1	6	TB5-21	6	18	BLACK	18.0
54	107F20	A4-1	2	A3-V1	2	15	BLACK	10.0
55	107G20	A3-V1	2	R1-R	TIN	15	BLACK	43.0
56	107H16	A7-L	6	TB5-21	6	18	BLACK	24.0
57	107J16	A7-L	6	K5-1	6	18	BLACK	21.0
58	NOT USED							
59	107M16	J5-37	24	TB5-21	6	18	BLACK	24.0
60	107N20	R1-R	TIN	R1-C	TIN	15	BLACK	4.0
61	108B20	J5-8	23	A4-2	2	15	BLACK	37.5
62	108C20	A4-2	2	A3-V2	2	15	BLACK	10.0
63	109C20	J5-9	23	TB5-22	2	15	BLACK	24.0
64	109D20	K11-2	2	TB5-22	2	15	BLACK	29.0
65	109E20	A4-3	2	TB5-22	2	15	BLACK	19.0
66	109F20	A4-3	2	A3-V3	2	15	BLACK	10.0
67	109G20	A1-4	2	TB5-22	2	15	BLACK	25.0
68	NOT USED							
69	110E16	J5-10	24	TB4-15	6	18	BLACK	16.0
70	110F16	J5-11	24	TB4-15	6	18	BLACK	17.0
71	110G20	S6-12	2	TB4-14	2	15	BLACK	29.5
72	110H20	A4-8	2	TB4-14	2	15	BLACK	20.5
73	110J20	A4-8	2	A3-N1	2	15	BLACK	18.0
74	110K20	A4-9	2	A3-N2	2	15	BLACK	16.5
75	110L20	A4-10	2	A3-N3	2	15	BLACK	15.0
76	110M20	A7-L	2	TB4-14	2	15	BLACK	20.0
77	110N20	K5-6	2	TB4-15	2	15	BLACK	20.0
78	NOT USED							
79	NOT USED							
80	111B20	J5-12	23	K8-1	2	15	BLACK	19.0
81	111C20	K8-1	2	A4-4	2	15	BLACK	29.0
82	111D20	A4-4	2	R10-1	TIN	15	BLACK	25.0
83	111B20	J5-13	23	A3-L1	2	15	BLACK	41.5
84	113B20	J5-14	23	K8-2	2	15	BLACK	18.0
85	113C20	K8-2	2	A4-5	2	15	BLACK	29.0
86	113D20	A4-5	2	R11-1	TIN	15	BLACK	25.5
87	114B20	J5-15	23	A3-L2	2	15	BLACK	39.0
88	115B20	J5-16	23	K3-8	2	15	BLACK	19.5
89	115C20	K8-3	2	A4-6	2	15	BLACK	29.0
90	115D20	A4-6	2	R12-1	TIN	15	BLACK	23.0
91	116B20	J5-17	23	A3-L3	2	15	BLACK	41.0
92	117A20	A3-S1	2	S6-16	2	15	BLACK	38.0

Figure 8. Control Box Harness Assembly (Sheet 18 of 25).

WIRE REF NO.	WIRE MARKING NO.	FROM	FIG.8 ITEM NO.	TO	FIG.8 ITEM NO.	FIG.8 ITEM NO.	MARKING COLOR	WIRE LENGTH
93	118A20	A3-S2	2	S6-20	2	15	BLACK	38.0
94	119A20	A3-S3	2	S6-24	2	15	BLACK	37.0
95	120A20	A3 (+)	2	M9 (+)	3	15	BLACK	34.0
96	NOT USED							
97	122C20	J5-19	23	TB5-13	2	15	BLACK	25.0
98	NOT USED							
99	122E20	R9-1	TIN	TB5-13	2	15	BLACK	22.0
100	123C20	J5-20	23	TB5-11	2	15	BLACK	27.5
101	NOT USED							
102	123E20	R7-1	TIN	TB5-11	2	15	BLACK	24.0
103	124A20	A3 (–)	2	M9 (–)	3	15	BLACK	38.0
104	125B20	J6-23	23	J3-X	–	15	RED	64.5
105	126A20	S16-3	2	K14-4	2	15	RED	48.0
106	127A20	A5-20	2	K15-7	2	15	RED	32.0
107	128A20	K16-2	2	K15-1	2	15	RED	20.0
108	129B20	S1-3	2	TB4-10	2	15	RED	27.0
109	129C20	TB4-10	2	M5-1	4	15	RED	15.0
110	129E20	M5-1	4	M6-1	4	15	RED	7.0
111	129F20	M6-1	4	M7-1	4	15	RED	11.0
112	129G20	R15-1	TIN	TB4-12	2	15	RED	30.0
113	129H20	K12-8	2	TB4-12	2	15	RED	18.5
114	129J20	K19-A	2	TB4-11	2	15	RED	19.0
115	129K20	P4-15	34	TB4-11	2	15	RED	26.0
116	129L20	P4-14	34	TB4-11	2	15	RED	26.0
117	129N20	K10-3	2	TB4-12	2	15	RED	14.5
118	129P20	TB4-12	2	K9-3	2	15	RED	11.5
119	129S20	CR6-1	TIN	TB4-10	2	15	RED	7.0
120	NOT USED							
121	131A20	J5-22	23	A1-10	2	15	BLACK	31.0
122	132A20	J5-23	23	A1-6	2	15	BLACK	31.0
123	133A20	J5-24	23	A1-12	2	15	BLACK	29.0
124	134A20	A1-8	2	K17-7	2	15	BLACK	33.0
125	135A20	A1-9	2	TB5-4	2	15	BLACK	34.5
126	135B20	R5-R	TIN	TB5-4	2	15	BLACK	33.5
127	135C20	J5-25	23	TB5-4	2	15	BLACK	31.0
128	135D20	S11-5	2	TB5-4	2	15	BLACK	26.0
129	136A20	S12-5	2	R1-L	TIN	15	BLACK	45.0
130	137A20	S12-4	2	A1-5	2	15	BLACK	34.0
131	NOT USED							
132	NOT USED							
128	135D20	S11-5	2	TB5-4	2	15	BLACK	26.0
129	136A20	S12-5	2	R1-L	TIN	15	BLACK	45.0
130	137A20	S12-4	2	A1-5	2	15	BLACK	34.0
131	NOT USED							
132	NOT USED							

Figure 8. Control Box Harness Assembly (Sheet 19 of 25).

WIRE REF NO.	WIRE MARKING NO.	FROM	FIG.8 ITEM NO.	TO	FIG.8 ITEM NO.	FIG.8 ITEM NO.	MARKING COLOR	WIRE LENGTH
133	138A20	J5-26	23	A1-7	2	15	BLACK	27.0
134	139A20	J5-27	23	A1-11	2	15	BLACK	28.0
135	140A20	A1-3	2	TB4-13	2	15	BLACK	19.0
136	140B20	J3-J	–	TB4-13	2	15	BLACK	45.0
137	140C20	J5-28	23	TB4-13	2	15	BLACK	17.0
138	140F20	K15-6	2	TB4-13	2	15	BLACK	42.5
139	141A20	A1-1	2	TB4-16	2	15	BLACK	18.0
140	141B20	J3-F	–	TB4-16	2	15	BLACK	47.0
141	141C20	J5-29	23	TB4-16	2	15	BLACK	16.0
142	141E20	CR2-2	TIN	TB4-16	2	15	BLACK	40.0
143	142B20	J5-30	23	TB6-12	2	15	BLACK	31.0
144	143C20	R5-C	TIN	TB4-21	2	15	BLACK	15.0
145	144C20	K17-4	2	TB4-22	2	15	BLACK	26.0
146	145C20	J3-G	–	TB4-18	2	15	BLACK	48.0
147	146C20	J3-C	–	TB4-19	2	15	BLACK	48.0
148	154A20	S12-2	2	A5-1	2	15	RED	31.0
149	155A20	A5-5	2	K21-4	2	15	RED	21.0
150	158A20	A5-12	2	A4-12	2	15	BLACK	36.0
151	158B20	K9-1	2	A4-12	2	15	BLACK	15.0
152	158C20	K9-1	2	R4-R	TIN	15	BLACK	25.0
153	159A20	A5-11	2	R4-C	TIN	15	BLACK	24.0
154	160A20	K15-8	2	A5-18	2	15	RED	31.0
155	161A20	A4-11	2	K9-2	2	15	BLACK	14.0
156	161B20	K9-2	2	R4-L	TIN	15	BLACK	25.0
157	162A20	CR1-2	TIN	S17-1	2	15	RED	45.0
158	164B20	J3-d	–	TB4-2	2	15	RED	41.0
159	166C20	J6-22	23	J3-N	–	15	RED	64.0
160	167C20	K15-5	2	TB5-5	2	15	RED	17.0
161	167D20	J6-21	23	TB5-5	2	15	RED	34.0
162	167E20	J3-T	–	TB5-5	2	15	RED	31.5
163	168D20	J6-5	23	TB5-7	2	15	RED	33.5
164	168E20	J3-b	–	TB5-7	2	15	RED	32.5
165	168F20	K14-8	2	TB5-7	2	15	RED	11.0
166	169B20	J6-20	23	TB5-8	2	15	RED	33.0
167	169C20	K16-7	2	TB5-8	2	15	RED	11.0
168	169D20	J3-a	–	TB5-8	2	15	RED	34.0
169	171A20	K19-B	2	TB5-6	2	15	RED	15.5
170	171B20	J6-19	23	TB5-6	2	15	RED	33.0
171	171C20	J3-Z	–	TB5-6	2	15	RED	32.0
172	172A20	S1-1	2	TB5-9	2	15	RED	26.0
173	172B20	J3-Y	–	TB5-9	2	15	RED	35.0
174	172C20	J6-18	23	TB5-9	2	15	RED	32.0
175	173A20	J5-31	23	S1-5	2	15	RED	44.0
176	174D20	J5-32	23	J3-0	–	15	RED	61.0
177	175A20	J5-33	23	M5-S	4	15	RED	50.0

Figure 8. Control Box Harness Assembly (Sheet 20 of 25).

WIRE REF NO.	WIRE MARKING NO.	FROM	FIG.8 ITEM NO.	TO	FIG.8 ITEM NO.	FIG.8 ITEM NO.	MARKING COLOR	WIRE LENGTH
178	176A20	J5-34	23	M6-S	4	15	RED	47.0
179	177A20	J5-35	23	M7-S	4	15	RED	44.0
180	179A20	MT4-2	4	M4 (+)	3	15	RED	37.0
181	180A20	CB1-1	4	TB5-2	2	15	RED	35.0
182	180B20	K14-A	2	TB5-1	2	15	RED	13.5
183	180C20	K14-9	2	TB5-1	2	15	RED	12.5
184	180D20	S1-8	2	TB5-1	2	15	RED	32.0
185	180E20	P4-4	34	TB5-1	2	15	RED	17.0
186	180F20	S2-2	2	S1-4	2	15	RED	29.0
187	NOT USED							
188	180H20	J3-U	–	TB5-2	2	15	RED	34.0
189	181A20	M2 (+)	3	A7 (+)	2	15	BLACK	34.0
190	182A20	M2 (–)	3	A7 (–)	2	15	BLACK	34.0
191	183A20	M8-1	3	S6-15	2	15	BLACK	11.0
192	184A20	M8-2	3	S6-17	2	15	BLACK	7.0
193	184B20	S6-13	2	K8-4	2	15	BLACK	31.0
194	184C20	K8-4	2	A4-7	2	15	BLACK	22.0
195	184D20	A4-7	2	R12-2	TIN	15	BLACK	32.0
196	185A20	K21-A	2	TB5-14	2	15	RED	14.0
197	185B20	CR4-2	TIN	TB5-15	2	15	RED	23.0
198	185C20	K21-6	2	TB5-14	2	15	RED	14.0
199	185D20	S12-3	2	TB5-15	2	17	RED	TBD
200	185E20	TB5-14	2	TB5-20	2	15	RED	6.5
201	185F20	A5-19	2	TB5-15	2	15	RED	31.0
202	185G20	TB5-15	2	M3 (+)	2	15	RED	TBD
203	186A20	M1 (–)	3	S6-7	2	15	BLACK	11.0
204	187A20	M1 (+)	3	S6-1	2	15	BLACK	15.0
205	188A20	S2-3	2	TB6-7	2	15	RED	32.0
206	188B20	DS1-1	2	TB6-7	2	15	RED	37.0
207	188C20	DS2-1	2	TB6-7	2	15	RED	42.0
208	188D20	DS3-1	2	TB6-7	2	15	RED	51.0
209	189A20	K15-B	2	CR8-1	TIN	15	RED	15.0
210	190A20	S1-7	2	TB6-8	2	15	RED	33.0
211	190B20	K15-A	2	TB6-8	2	15	RED	29.5
212	190C20	CR4-1	TIN	TB6-8	2	15	RED	9.0
213	190E20	J3-P	–	K15-A	2	15	RED	55.5
214	191A16	J5-18	24	J1-2	11	18	BLACK	52.5
215	NOT USED							
216	193A16	J5-21	24	J1-3	11	18	BLACK	53.0
217	194A20	K11-3	2	R7-2	TIN	15	BLACK	20.0
218	195A20	DS4-2	2	S11-12	2	15	BLACK	12.0
219	196A20	DS5-2	2	K10-2	2	15	BLACK	20.0
220	196B20	K10-2	2	S11-9	2	15	BLACK	16.0
221	197A20	S11-11	2	TB5-10	2	15	BLACK	23.0
222	197B20	R6-1	TIN	TB5-10	2	15	BLACK	24.0

Figure 8. Control Box Harness Assembly (Sheet 21 of 25).

WIRE REF NO.	WIRE MARKING NO.	FROM	FIG.8 ITEM NO.	TO	FIG.8 ITEM NO.	FIG.8 ITEM NO.	MARKING COLOR	WIRE LENGTH
223	197C20	K11-1	2	TB5-10	2	15	BLACK	11.0
224	198A20	S11-8	2	TB5-12	2	15	BLACK	22.0
225	198B20	R8-1	TIN	TB5-12	2	15	BLACK	23.0
226	198C20	K11-8	2	TB5-12	2	15	BLACK	12.5
227	199A20	S5-3	2	K8-11	2	15	RED	20.0
228	200A20	K11-6	2	R9-2	TIN	15	BLACK	18.5
229	201A20	MT4-3	4	M4 (–)	3	15	RED	39.0
230	202A20	P4-3	34	K12-9	2	15	RED	39.0
231	203A20	P4-9	34	K5-3	2	15	RED	23.0
232	204A20	P4-10	34	K16-9	2	15	RED	43.5
233	205A20	P4-11	34	K8-9	2	15	RED	31.0
234	206A20	P4-13	34	S7-4	2	15	RED	40.0
235	207A20	CB1-2	4	S17-2	2	15	RED	53.0
236	208A20	P4-17	34	K9-6	2	15	RED	11.0
237	209A20	P4-18	34	K8-6	2	15	RED	27.0
238	168G20	TB5-7	2	K8-5	2	15	RED	17.0
239	168H20	K5-2	2	K19-6	2	15	RED	42.0
240	168J20	K5-2	2	K8-10	2	15	RED	19.0
241	168K20	K8-10	2	K9-5	2	15	RED	24.0
242	168L20	K9-5	2	K5-7	2	15	RED	21.5
243	168M20	K5-7	2	K8-5	2	15	RED	19.0
244	212A20	K12-3	2	K13-3	2	15	RED	4.5
245	NOT USED							
246	214A20	P4-1	34	J16-4	2	15	RED	41.0
247	215A20	P4-7	34	K14-6	2	15	RED	40.0
248	NOT USED							
249	NOT USED							
250	217A20	K16-6	2	K5-8	2	15	RED	43.0
251	NOT USED							
252	220A20	P4-2	34	K19-9	2	15	RED	46.0
253	221A20	S7-5	2	TB5-16	2	15	RED	26.0
254	NOT USED							
255	221C20	K12-A	2	TB5-16	2	15	RED	11.0
256	221D20	K12-5	2	TB5-16	2	15	RED	14.5
257	NOT USED							
258	223A20	J6-6	23	P4-8	34	15	RED	47.0
259	NOT USED							
260	NOT USED							
261	225E20	J3-S	–	TB5-17	2	15	RED	36.5
262	NOT USED							
263	225G20	K21-9	2	TB5-17	2	15	RED	11.5
264	225H20	K21-9	2	K12-2	2	15	RED	24.0
265	228A20	S7-7	2	K12-7	2	15	RED	59.0
266	NOT USED							
267	NOT USED							

Figure 8. Control Box Harness Assembly (Sheet 22 of 25).

WIRE REF NO.	WIRE MARKING NO.	FROM	FIG.8 ITEM NO.	TO	FIG.8 ITEM NO.	FIG.8 ITEM NO.	MARKING COLOR	WIRE LENGTH
268	NOT USED							
269	NOT USED							
270	233A20	K14-B	2	TB5-19	2	15	RED	14.5
271	233B20	K21-7	2	TB5-19	2	15	RED	12.0
272	233C20	S16-2	2	TB5-19	2	15	RED	22.0
273	234A20	K15-4	2	R14-1	TIN	15	RED	21.0
274	239A20	DS7-1	TIN	TB6-9	2	15	RED	30.0
275	239B20	R15-2	TIN	TB6-9	2	15	RED	9.5
276	239C20	DS6-1	TIN	TB6-9	2	15	RED	36.0
277	240B20	J5-7	23	TB5-18	2	15	RED	27.0
278	240C20	K17-B	2	TB5-18	2	15	RED	8.5
279	240D20	S5-5	2	TB5-18	2	15	RED	19.5
280	240E20	CR3-1	TIN	TB5-18	2	15	RED	18.0
281	241A20	S7-9	2	DS6-2	TIN	15	RED	8.0
282	242A20	S5-2	2	CR3-2	TIN	15	RED	42.5
283	129R20	K12-8	2	K16-5	2	15	RED	11.0
284	248C20	CR6-2	TIN	TB4-3	2	15	RED	7.0
285	249A20	K16-A	2	K16-8	2	15	RED	3.0
286	134B20	K17-7	2	S11-4	2	15	BLACK	52.0
287	135F20	A1-9	2	S11-2	2	15	BLACK	51.0
288	142A20	TB6-12	2	R5-L	TIN	15	BLACK	33.0
289	142D20	TB6-12	2	S11-1	2	15	BLACK	28.0
290	250A20	K12-1	2	CR8-2	TIN	15	RED	3.0
291	128B20	K16-2	2	CR7-1	TIN	15	RED	6.0
292	249B20	K16-8	2	CR7-2	TIN	15	RED	6.0
293	100BD20	J3-D	–	K12-B	2	15	RED	50.0
294	301A20	K13-7	2	K8-12	2	15	RED	47.0
295	302A20	P4-16	34	S7-1	2	15	RED	42.0
296	303A20	K13-A	2	S7-2	2	15	RED	63.0
297	305A20	K13-9	2	K14-2	2	15	RED	19.5
298	306A20	S5-6	2	K10-8	2	15	RED	27.0
299	NOT USED							
				USABLE ON 86Q, 86R ONLY				
300	225N16	J61-1	206	TB5-17	6	18	RED	45.0
301	164N16	J61-2	206	CB2-2	11	18	RED	30.0
302	165R16	J61-3	206	MT4-4	4A	18	RED	28.0
303	163K16	MT4-1	4A	CB2-1	11	18	RED	21.0

Figure 8. Control Box Harness Assembly (Sheet 23 of 25).

SHIELDED CABLE CONNECTION TABLE						
CABLE ITEM NO.	WIRE	FROM	FIG.8 ITEM NO.	TO	FIG.8 ITEM NO.	WIRE LENGTH
38	(SHIELD)	(CUTOFF)	-	A5-15	40	27.5
	156A20	K17-5	40	A5-14	40	27.5
	157A20	K17-6	40	A5-13	40	27.5
45	(SHIELD)	(CUTOFF)	-	TB4-20	40	56.0
	143B20	J3-H	40	TB4-21	40	56.0
	144B20	J3-E	40	TB4-22	40	56.0
52	(SHIELD)	(CUTOFF)	-	TB4-20	40	17.0
	145A20	K17-8	40	TB4-18	40	17.0
	146A20	K17-9	40	TB4-19	40	17.0
59	(SHIELD)	(CUTOFF)	-	TB4-4	64	26.0
	163C16	J6-7	61	TB4-1	62	26.0
67	(SHIELD)	CBGND	72	(CUTOFF)	-	26.0
	100J16	CBGND	69	A5-4	71	26.0
75	(SHIELD)	K15-9	79	(CUTOFF)	-	27.0
	151A16	K21-5	77	A5-3	77	27.0
82	(SHIELD)	(CUTOFF)	-	A5-GP	86	30.0
	165G16	MT4-4	87	K21-8	84	30.0
90	(SHIELD)	(CUTOFF)	-	TB4-5	94	39.0
	178C16	CR1-1	-	TB4-6	92	39.0
97	(SHIELD)	(CUTOFF)	-	TB4-4	101	13.5
	163A16	MT4-1	102	TB4-1	99	13.5
105	(SHIELD)	J3-GND	109	(CUTOFF)	-	55.0
	151B16	J3-M	107	K21-5	-	55.0
112	(SHIELD)	(CUTOFF)	-	TB4-20	116	46.0
	145B16	J2-A	-	TB4-18	114	46.0
	146B16	J2-B	-	TB4-19	114	46.0
120	(SHIELD)	(CUTOFF)	-	TB4-9	124	44.0
	147D16	J3-y	-	TB4-7	122	44.0
	148D16	J3-x	-	TB4-8	122	44.0
128	(SHIELD)	(CUTOFF)	-	CBGND	134	20.0
	100E16	J6-1	131	CBGND	132	20.0
	165D16	J6-2	131	MT4-4	133	20.0
138	(SHIELD)	(CUTOFF)	-	TB4-9	142	12.0
	147C16	A5-16	140	TB4-7	140	12.0
	148C16	A5-17	140	TB4-8	140	12.0
146	(SHIELD)	(CUTOFF)	-	TB4-4	150	19.5
UOC: EMK,	163B16	FU1-1	-	TB4-1	148	19.5
YNN	164A16	FU1-2	-	TB4-2	148	19.5
154	(SHIELD)	K2-GND	157	TB4-20	159	46.0
	143A16	J2-C	-	TB4-21	158	46.0
	144A16	J2-D	-	TB4-22	158	46.0

Figure 8. Control Box Harness Assembly (Sheet 24 of 25).

SHIELDED CABLE CONNECTION TABLE						
CABLE ITEM NO.	WIRE	FROM	FIG.8 ITEM NO.	TO	FIG.8 ITEM NO.	WIRE LENGTH
164	(SHIELD)	J6-15	169	A5-GP	170	35.0
	149A16	J6-13	168	A5-10	167	35.0
	150A16	J6-14	168	A5-9	167	35.0
174	(SHIELD)	J6-12	180	TB4-9	178	23.0
	147A16	J6-10	179	TB4-7	177	23.0
	148A16	J6-11	179	TB4-8	177	23.0
184	(SHIELD)	A5-GP	188	(CUTOFF)	-	53.0
	152A20	A5-8	186	R2-L	-	53.0
	153A20	A5-6	186	R2-R	-	53.0
	192B20	A5-7	186	R2-C	-	53.0
193	(SHIELD)	J6-9	196	TB4-5	199	24.5
	(SHIELD)	(CUTOFF)	-	TB4-5	201	24.5
	164C16	J6-4	196	TB4-2	199	24.5
	178B16	J6-8	196	TB4-6	199	24.5
	248B16	J6-3	196	TB4-3	200	24.5
208	(SHIELD)	(CUTOFF)	-	TB4-4	212	19.5
UOC: 86Q,	163B16	CB2-1	216	TB4-1	210	19.5
86R	164A16	CB2-2	-	TB4-2	210	19.5

Figure 8. Control Box Harness Assembly (Sheet 25 of 25).

(1)	(2) SMR CODE a. ARMY	b. AIR FORCE	(3) NSN	(4) CAGEC	(5) PART NUMBER	(6) DESCRIPTION AND USABLE ON CODE (UOC)	(7) QTY
ITEM NO							
						GROUP 03 CONTROL BOX ASSEMBLY	
						FIG. 8 CONTROL BOX HARNESS ASSEMBLY	
1	PBFFF	PBOOO	6150-01-383-6511	30554	88-22303	. . WIRING HARNESS, BRANCH (NOT SHOWN) UOC: EMK	1
1	PBFFF	PBOOO	6150-01-384-0013	30554	88-22664	. . WIRING HARNESS, BRANCH (NOT SHOWN) UOC: YNN	1
1	PBFFF	PBOOO		30554	97-24049	. . WIRING HARNESS, BRANCH (NOT SHOWN) UOC: 86Q	1
1	PBFFF	PBOOO		30554	97-24060	. . WIRING HARNESS, BRANCH (NOT SHOWN) UOC: 86R	1
2	PAFZZ	PAOZZ	5940-01-425-2020	30554	88-20274-1	...TERMINAL, SPADE #6, 22-18 AWG UOC: EMK, 86Q	366
2	PAFZZ	PAOZZ	5940-01-425-2020	30554	88-20274-1	...TERMINAL, SPADE #6 22-18 AWG UOC: YNN, 86R	360
3	PAFZZ	PAOZZ	5940-00-113-8184	96906	MS25036-150	...TERMINAL, LUG 1/4, 22-18 AWG	10
4	PAFZZ	PAOZZ	5940-01-531-6448	30554	88-20274-3	...TERMINAL, SPADE #8, 22-18 AWG	18
4	PAFZZ	PAOZZ	5940-00-283-5281	96906	MS25036-109	...TERMINAL, LUG 5/16, 16-14 AWG UOC: 86Q, 86R	2
5	PAFZZ	PAOZZ	5940-01-369-2270	30554	88-20274-4	...TERMINAL, SPADE #10, 22-18 AWG	1
6	PAFZZ	PAOZZ	5940-01-110-6423	30554	88-20274-5	...TERMINAL, SPADE #6, 16-14 AWG UOC: EMK, YNN	13
6	PAFZZ	PAOZZ	5940-01-367-9569	30554	88-20274-5	...TERMINAL, SPADE #6, 16-14 AWG UOC: 86Q, 86R	14
7	PAFZZ	PAOZZ	5940-01-367-9569	30554	88-20274-6	...TERMINAL, SPADE #10, 16-14 AWG	1
8	PAFZZ	PAOZZ	5940-00-557-1629	96906	MS25036-149	...TERMINAL, LUG #8,22-18 AWG	1
9	PAFZZ	PAOZZ	5940-00-113-9828	96906	MS25036-148	...TERMINAL, LUG #4,22-18 AWG	1
10	PAFZZ	PAOZZ	5940-01-368-6774	30554	88-22071	...TERMINAL, LUG #4,12-10 AWG	1
11	PAFZZ	PAOZZ	5940-01-259-2190	30554	88-20274-11	...TERMINAL, SPADE #8, 16-14 AWG	3
12	PAFZZ	PAOZZ	5975-00-111-3208	96906	MS3367-5-9	...STRAP, TIEDOWN,ELECT	V
13	PAFZZ	PAOZZ	5975-00-074-2072	96906	MS3367-1-9	...STRAP, TIEDOWN,ELECT	V
14	PAFZZ	PAOZZ	5975-00-944-1499	96906	MS3368-1-9A	...STRAP, TIEDOWN,ELECT	1
15	MFFZZ	MOO	6145-00-851-8505	81349	M5086/2-20-9	. . . WIRE, ELECTRICAL	1
16	XDFZZ	XB	5999-01-366-2621	30554	88-21183-2	...CONTACT,ELECTRICAL	11
17	XDFZZ	XB	6150-01-384-0013	30554	88-21181-12	...TERMINALBLOCK TB6	1
18	MFFZZ	MOO	6145-00-578-6605	81349	M5086/2-16-9	. . . WIRE, ELECTRICAL	1

(1)	(2) SMR CODE a.	b.	(3)	(4)	(5)	(6)	(7)
ITEM NO	ARMY	AIR FORCE	NSN	CAGEC	PART NUMBER	DESCRIPTION AND USABLE ON CODE (UOC)	QTY

19 XDFZZ XB 5940-01-531-6040 30554 88-21181-22 ...TERMINALBLOCK TB4AND
TB5 2

20 PAFZZ PAOZZ 5310-00-836-3520 30554 69-561-1 . . . NUT, PLAIN, ASSEMBLED 4
21 PAFZZ PAOZZ 5305-01-367-2314 30554 69-662-11 ...SCREW, ASSEMBLED WAS 4
22 PAFZZ PAOZZ 5935-01-312-7038 30554 88-22312 . . . CLAMP, CABLE, ELECTRICAL 2
23 PAFZZ PAOZZ 5999-01-366-5119 30554 88-21941 ...PIN 41
24 PAFZA PAOZA 5999-01-463-4017 30554 88-31943 ...PIN .... 7 25 PAFZZ PAOZZ 5935-01-014-7861
30554 88-21939 ...CONNECTOR BODY,RECEP,
J5 1

26 PAFZZ PAOZZ 5935-01-104-6137 30554 88-22313 . . . EXTENSION, ELECTRICAL .... 1 27 PAFZZ PAOZZ 5935-01-102-
0366 30554 88-21981 ...CONNECTOR BODY,RECEP,
J6 1

28 PAFZZ PAOZZ 5935-01-367-4422 30554 88-21084 ...SOCKET,PLUG-IN, ELECT
K12-K21 8

29 PAFZZ PAOZZ 5935-01-383-2612 30554 88-21897-1 ...SOCKET,PLUG-IN, ELECT,
K11 1

30 PAFZZ PAOZZ 5935-00-823-5322 96906 MS25251-16 ...PLUG,END SEAL, ELECT 29
31 PAFZZ PAOZZ 5935-01-102-7124 19207 12258941 ...CONNECTOR,RECEP, J3 1
32 PAFZZ PAOZZ 5935-00-852-9611 96906 MS3102R18-11P ...PLUG,CONNECT,ELECT,J2 1
33 MFFZZ MOO 6145-00-578-7514 81349 M5086/2-12-9 . . WIRE, ELECTRICAL 1
34 PAFZZ PAOZZ 5999-01-320-7853 30554 88-21730 ...CONTACT,ELECTRICAL 17
35 PAFZZ PAOZZ 5935-01-366-9934 30554 88-21729 ...CONNECTOR BODY,PLUG .... 1 36 PAFZZ PAOZZ 5961-00-295-5757
50434 1901-0759 . . . SEMICONDUCTOR DEVICE CR6,
CR7, AND CR8 3

37 MFFZZ MOO 5970-00-812-2969 81349 M23053/5-104-0 . . INSULATION, SLEEVING 1.25
INCHES REQUIRED 4

38 AFFFF AOO 30554 88-22098-1 ...CABLE ASSY, SHIELDED (NOT
SHOWN) 1

39 MFFZZ MOO 30554 88-22098-1-1 ....CABLE, SHIELDED MAKE FROM
P/N M27500-20TE2T14 (81349), 27.5
INCHES REQUIRED 1

40 PAFZZ PAOZZ 5940-01-425-2020 30554 88-20274-1 ....TERMINAL, SPADE #6, 22-18
AWG 5

41 PAFZZ PAOZZ 5940-01-135-7081 30554 88-20596-2 ....SHIELD, TERMINATION CUT
WIRE LEAD TO 3 INCHES 1

42 MFFZZ MOO 30554 88-22629-004 ....MARKER,IDENTIFICATION
MAKE FROM P/N M23053/5-105-9
(81349), 1.25 INCHES REQUIRED 1

43 MFFZZ MOO 30554 88-22629-002 ....MARKER,IDENTIFICATION
MAKE FROM P/N M23053/5-105-9
(81349), 1.25 INCHES REQUIRED 2

44 MFFZZ MOO 30554 88-22629-005 ....MARKER,IDENTIFICATION
MAKE FROM P/N M23053/5-105-9
(81349), 1.25 INCHES REQUIRED 1

45 AFFFF AOO 30554 88-22098-2 ...CABLE ASSY, SHIELDED (NOT
SHOWN) 1

46 MFFZZ MOO 30554 88-22098-2-1 ....CABLE, SHIELDED MAKE FROM
P/N M27500-20TE2T14 (81349), 56
INCHES REQUIRED 1

47 PAFZZ PAOZZ 5940-01-425-2020 30554 88-20274-1 ....TERMINAL, SPADE #6, 22-18
AWG 3

48 PAFZZ PAOZZ 5940-01-135-7081 30554 88-20596-2 ....SHIELD, TERMINATION CUT
WIRE LEAD TO 4 INCHES. 1

49 MFFZZ MOO 30554 88-22629-006 ....MARKER,IDENTIFICATION
MAKE FROM P/N M23053/5-105-9
(81349),1.25 INCHES REQUIRED 1

(1) (2) (3) (4) (5) (6) (7)
SMR CODE
a. b.
ITEM NO ARMY AIR FORCE NSN CAGEC PART NUMBER DESCRIPTION AND USABLE ON CODE (UOC) QTY

50 MFFZZ MOO 30554 88-22629-007 ....MARKER,IDENTIFICATION
MAKE FROM P/N M23053/5-105-9
(81349), 1.25 INCHES REQUIRED 2
51 MFFZZ MOO 30554 88-22629-008 ....MARKER,IDENTIFICATION
MAKE FROM P/N M23053/5-105-9
(81349), 1.25 INCHES REQUIRED 2
52 AFFFF AOO 30554 88-22098-3 ...CABLE ASSY, SHIELDED (NOT
SHOWN) 1
53 MFFZZ MOO 30554 88-22098-3-1 ....CABLE, SHIELDED MAKE FROM
P/N M27500-20TE2T14 (81349), 17
INCHES REQUIRED 1
54 PAFZZ PAOZZ 5940-01-425-2020 30554 88-20274-1 ....TERMINAL, SPADE #6, 22-18
AWG 5
55 PAFZZ PAOZZ 5970-00-812-2969 30554 88-20596-2 ....SHIELD, TERMINATION CUT
WIRE LEAD TO 4 INCHES 1
56 MFFZZ MOO 30554 88-22629-009 ....MARKER,IDENTIFICATION
MAKE FROM P/N M23053/5-105-9
(81349), 1.25 INCHES REQUIRED 1
57 MFFZZ MOO 30554 88-22629-010 ....MARKER,IDENTIFICATION
MAKE FROM P/N M23053/5-105-9
(81349), 1.25 INCHES REQUIRED 2
58 MFFZZ MOO 30554 88-22629-011 ....MARKER,IDENTIFICATION
MAKE FROM P/N M23053/5-105-9
(81349), 1.25 INCHES REQUIRED 2
59 AFFFF AOO 30554 88-22099-1 ...CABLE ASSY, SHIELDED (NOT
SHOWN) 1
60 MFFZZ MOO 30554 88-22099-1-1 ....CABLE, SHIELDED MAKE FROM
P/N M27500-16TE1T14 (81349), 26
INCHES REQUIRED 1
61 PAFZA PAOZA 5999-01-463-4017 30554 88-21943 ....PIN .... 1 62 PAFZZ PAOZZ 5940-01-110-6423
30554 88-20274-5 ....TERMINAL, SPADE #6, 16-14
AWG 1
63 PAFZZ PAOZZ 5940-01-531-5519 30554 88-20596-1 ....SHIELD, TERMINATION CUT
WIRE LEAD TO 4 INCHES 1
64 PAFZZ PAOZZ 5940-01-425-2020 30554 88-20274-1 ....TERMINAL, SPADE #6, 22-18
AWG 1
65 MFFZZ MOO 30554 88-22629-012 ....MARKER,IDENTIFICATION
MAKE FROM P/N M23053/5-105-9
(81349), 1.25 INCHES REQUIRED 1
66 MFFZZ MOO 30554 88-22629-013 ....MARKER,IDENTIFICATION
MAKE FROM P/N M23053/5-105-9
(81349), 1.25 INCHES REQUIRED 2
67 AFFFF AOO 30554 88-22099-2 ...CABLE ASSY, SHIELDED (NOT
SHOWN) 1
68 MFFZZ MOO 30554 88-22099-2-1 ....CABLE, SHIELDED MAKE FROM
P/N M27500-16TE1T14 (81349), 26
INCHES REQUIRED 1
69 PAFZZ PAOZZ 5940-01-110-6423 30554 88-20274-5 ....TERMINAL, SPADE #6, 16-14
AWG 1
70 PAFZZ PAOZZ 5940-01-531-5519 30554 88-20596-1 ....SHIELD, TERMINATION CUT
WIRE LEAD TO 2 INCHES 1
71 PAFZZ PAOZZ 5940-01-367-9569 30554 88-20274-6 ....TERMINAL, SPADE #10, 16-14
AWG 1
72 PAFZZ PAOZZ 5940-01-369-2270 30554 88-20274-4 ....TERMINAL, SPADE #10, 22-18
AWG 1

(1)	(2)		(3)	(4)	(5)	(6)	(7)
	SMR CODE						
	a.	b.					
ITEM NO	ARMY	AIR FORCE	NSN	CAGEC	PART NUMBER	DESCRIPTION AND USABLE ON CODE (UOC)	QTY

73 MFFZZ MOO 30554 88-22629-014 ....MARKER,IDENTIFICATION
MAKE FROM P/N M23053/5-105-9 (81349), 1.25 INCHES REQUIRED 1

74 MFFZZ MOO 30554 88-22629-015 ....MARKER,IDENTIFICATION
MAKE FROM P/N M23053/5-105-9 (81349), 1.25 INCHES REQUIRED 2

75 AFFFF AOO 30554 88-22099-3 ...CABLE ASSY, SHIELDED (NOT
SHOWN) 1

76 MFFZZ MOO 30554 88-22099-3-1 ....CABLE, SHIELDED MAKE FROM
P/N M27500-16TE1T14 (81349), 27 INCHES REQUIRED 1

77 PAFZZ PAOZZ 5940-01-110-6423 30554 88-20274-5 ....TERMINAL, SPADE #6, 16-14
AWG 2

78 PAFZZ PAOZZ 5940-01-531-5519 30554 88-20596-1 ....SHIELD, TERMINATION CUT
WIRE LEAD TO 3 INCHES 1

79 PAFZZ PAOZZ 5940-01-425-2020 30554 88-20274-1 ....TERMINAL, SPADE #6, 22-18
AWG 1

80 MFFZZ MOO 30554 88-22629-016 ....MARKER,IDENTIFICATION
MAKE FROM P/N M23053/5-105-9 (81349), 1.25 INCHES REQUIRED 1

81 MFFZZ MOO 30554 88-22629-017 ....MARKER,IDENTIFICATION
MAKE FROM P/N M23053/5-105-9 (81349), 1.25 INCHES REQUIRED 2

82 AFFFF AOO 30554 88-22099-7 ...CABLE ASSY, SHIELDED (NOT
SHOWN) 1

83 MFFZZ MOO 30554 88-22099-7-1 ....CABLE, SHIELDED MAKE FROM
P/N M27500-16TE1T14 (81349), 30 INCHES REQUIRED 1

84 PAFZZ PAOZZ 5940-01-110-6423 30554 88-20274-5 ....TERMINAL, SPADE #6, 16-14
AWG 1

85 PAFZZ PAOZZ 5940-01-531-5519 30554 88-20596-1 ....SHIELD, TERMINATION CUT
WIRE LEAD TO 6 INCHES 1

86 PAFZZ PAOZZ 5940-01-367-9569 30554 88-20274-6 ....TERMINAL, SPADE #10, 16-14
AWG 1

87 PAFZZ PAOZZ 5940-00-230-0515 96906 MS25036-154 ....TERMINAL, LUG .... 1 88 MFFZZ MOO 30554 88-22629-019 ....MARKER,IDENTIFICATION
MAKE FROM P/N M23053/5-105-9 (81349), 1.25 INCHES REQUIRED 2

89 MFFZZ MOO 30554 88-22629-024 ....MARKER,IDENTIFICATION
MAKE FROM P/N M23053/5-105-9 (81349), 1.25 INCHES REQUIRED 1

90 AFFFF AOO 30554 88-22099-5 ...CABLE ASSY, SHIELDED (NOT
SHOWN) 1

91 MFFZZ MOO 30554 88-22099-5-1 ....CABLE, SHIELDED MAKE FROM
P/N M27500-16TE1T14 (81349), 39 INCHES REQUIRED 1

92 PAFZZ PAOZZ 5940-01-110-6423 30554 88-20274-5 ....TERMINAL, SPADE #6, 16-14
AWG 1

93 PAFZZ PAOZZ 5940-01-531-5519 30554 88-20596-1 ....SHIELD, TERMINATION CUT
WIRE LEAD TO 4 INCHES 1

94 PAFZZ PAOZZ 5940-01-425-2020 30554 88-20274-1 ....TERMINAL, SPADE #6, 22-18
AWG 1

95 MFFZZ MOO 30554 88-22629-020 ....MARKER,IDENTIFICATION
MAKE FROM P/N M23053/5-105-9 (81349), 1.25 INCHES REQUIRED 1

(1)	(2) SMR CODE a.	b.	(3)	(4)	(5)	(6)	(7)
ITEM NO	ARMY	AIR FORCE	NSN	CAGEC	PART NUMBER	DESCRIPTION AND USABLE ON CODE (UOC)	QTY

96 MFFZZ MOO 30554 88-22629-021 ....MARKER IDENTIFICATION MAKE FROM P/N M23053/5-105-9 (81349), 1.25 INCHES REQUIRED 2

97 AFFFF AOO 30554 88-22099-6 ...CABLE ASSY, SHIELDED (NOT SHOWN) 1

98 MFFZZ MOO 30554 88-22099-6-1 ....CABLE, SHIELDED MAKE FROM P/N M27500-16TE1T14 (81349), 3.5 INCHES REQUIRED 1

99 PAFZZ PAOZZ 5940-01-110-6423 30554 88-20274-5 ....TERMINAL, SPADE #6, 16-14 AWG 1

100 MFFZZ MOO 5940-01-531-5519 30554 88-20596-1 ....SHIELD, TERMINATION CUT WIRE LEAD TO 4 INCHES 1

101 PAFZZ PAOZZ 5940-01-425-2020 30554 88-20274-1 ....TERMINAL, SPADE #6, 22-18 AWG 1

102 PAFZZ PAOZZ 5940-00-230-0515 96906 MS25036-154 ....TERMINAL, LUG .... 1 103 MFFZZ MOO 30554 88-22629-022 ....MARKER,IDENTIFICATION MAKE FROM P/N M23053/5-105-9 (81349), 1.25 INCHES REQUIRED 1

104 MFFZZ MOO 30554 88-22629-023 ....MARKER,IDENTIFICATION MAKE FROM P/N M23053/5-105-9 (81349), 1.25 INCHES REQUIRED 2

105 AFFFF AOO 30554 88-22100 ...CABLE ASSY, SHIELDED (NOT SHOWN) 1

106 MFFZZ MOO 30554 88-22100-1 ....CABLE, SHIELDED MAKE FROM P/N M27500-16TE1T14 (81349), 55 INCHES REQUIRED 1

107 PAFZZ PAOZZ 5940-01-110-6423 30554 88-20274-5 ....TERMINAL, SPADE #6, 16-14 AWG 1

108 PAFZZ PAOZZ 5940-01-531-5519 30554 88-20596-1 ....SHIELD, TERMINATION CUT WIRE LEAD TO 6 INCHES 1

109 PAFZZ PAOZZ 5940-00-557-1629 96906 MS25036-149 ....TERMINAL, LUG#8, 22-18 AWG 1

110 MFFZZ MOO 30554 88-22629-025 ....MARKER,IDENTIFICATION MAKE FROM P/N M23053/5-105-9 (81349), 1.25 INCHES REQUIRED 1

111 MFFZZ MOO 30554 88-22629-026 ....MARKER,IDENTIFICATION MAKE FROM P/N M23053/5-105-9 (81349), 1.25 INCHES REQUIRED 2

112 AFFFF AOO 30554 88-22101-1 ...CABLE ASSY, SHIELDED (NOT SHOWN) 1

113 MFFZZ MOO 30554 88-22101-1-1 ....CABLE, SHIELDED MAKE FROM P/N M27500-16TE2T14 (81349), 46 INCHES REQUIRED 1

114 PAFZZ PAOZZ 5940-01-110-6423 30554 88-20274-5 ....TERMINAL, SPADE #6, 16-14 AWG 2

115 PAFZZ PAOZZ 5940-01-531-5519 30554 88-20596-1 ....SHIELD, TERMINATION CUT WIRE LEAD TO 4 INCHES 1

116 PAFZZ PAOZZ 5940-01-425-2020 30554 88-20274-1 ....TERMINAL, SPADE #6, 22-18 AWG 1

117 MFFZZ MOO 30554 88-22629-027 ....MARKER,IDENTIFICATION MAKE FROM P/N M23053/5-105-9 (81349), 1.25 INCHES REQUIRED 1

118 MFFZZ MOO 30554 88-22629-028 ....MARKER,IDENTIFICATION MAKE FROM P/N M23053/5-105-9 (81349), 1.25 INCHES REQUIRED 2

(1)	(2)		(3)	(4)	(5)	(6)	(7)
	SMR CODE						
	a.	b.					
ITEM NO	ARMY	AIR FORCE	NSN	CAGEC	PART NUMBER	DESCRIPTION AND USABLE ON CODE (UOC)	QTY

119 MFFZZ MOO 30554 88-22629-029 ....MARKER,IDENTIFICATION
MAKE FROM P/N M23053/5-105-9
(81349), 1.25 INCHES REQUIRED 2

120 AFFFF AOO 30554 88-22101-2 ...CABLE ASSY, SHIELDED (NOT
SHOWN) 1

121 MFFZZ MOO 30554 88-22101-2-1 ....CABLE, SHIELDED MAKE FROM
P/N M27500-16TE2T14 (81349), 44
INCHES REQUIRED 1

122 PAFZZ PAOZZ 5940-01-110-6423 30544 88-20274-5 ....TERMINAL, SPADE #6, 16-14
AWG 2

123 PAFZZ PAOZZ 5940-01-531-5519 30554 88-20596-1 ....SHIELD, TERMINATION CUT
WIRE LEAD TO 4 INCHES 1

124 PAFZZ PAOZZ 5940-01-425-2020 30554 88-20274-1 ....TERMINAL, SPADE #6, 22-18
AWG 1

125 MFFZZ MOO 30554 88-22629-030 ....MARKER,IDENTIFICATION
MAKE FROM P/N M23053/5-105-9
(81349), 1.25 INCHES REQUIRED 1

126 MFFZZ MOO 30554 88-22629-031 ....MARKER,IDENTIFICATION
MAKE FROM P/N M23053/5-105-9
(81349), 1.25 INCHES REQUIRED 2

127 MFFZZ MOO 30554 88-22629-032 ....MARKER,IDENTIFICATION
MAKE FROM P/N M23053/5-105-9
(81349), 1.25 INCHES REQUIRED 2

128 AFFFF AOO 30554 88-22101-3 ...CABLE ASSY, SHIELDED (NOT
SHOWN) 1

129 MFFZZ MOO 30554 88-22101-3-1 ....CABLE, SHIELDED MAKE FROM
P/N M27500-16TE2T14 (81349), 20.0
INCHES REQUIRED 1

130 PAFZZ PAOZZ 5940-01-531-5519 30554 88-20596-1 ....SHIELD, TERMINATION CUT
WIRE LEAD TO 4 INCHES 1

131 PAFZA PAOZA 5999-01-463-4017 30554 88-21943 ....PIN .... 2 132 PAFZZ PAOZZ 5940-01-367-9569
30554 88-20274-6 ....TERMINAL, SPADE #10, 16-14
AWG 1

133 PAFZZ PAOZZ 5940-00-230-0515 96906 MS25036-154 ....TERMINAL, LUG .... 1 134 PAFZZ PAOZZ 5940-01-369-
2270 30554 88-20274-4 ....TERMINAL, SPADE #10, 22-18
AWG 1

135 MFFZZ MOO 30554 88-22629-033 ....MARKER,IDENTIFICATION
MAKE FROM P/N M23053/5-105-9
(81349), 1.25 INCHES REQUIRED 1

136 MFFZZ MOO 30554 88-22629-034 ....MARKER,IDENTIFICATION
MAKE FROM P/N M23053/5-105-9
(81349), 1.25 INCHES REQUIRED 2

137 MFFZZ MOO 30554 88-22629-035 ....MARKER,IDENTIFICATION
MAKE FROM P/N M23053/5-105-9
(81349), 1.25 INCHES REQUIRED 2

138 AFFFF AOO 30554 88-22101-4 ...CABLE ASSY, SHIELDED (NOT
SHOWN) 1

139 MFFZZ MOO 30554 88-22101-4-1 ....CABLE, SHIELDED MAKE FROM
P/N M27500-16TE2T14 (81349), 12
INCHES REQUIRED 1

140 PAFZZ PAOZZ 5940-01-110-6423 30554 88-20274-5 ....TERMINAL, SPADE #6, 16-14
AWG 4

141 PAFZZ PAOZZ 5940-01-531-5519 30554 88-20596-1 ....SHIELD, TERMINATION CUT
WIRE LEAD TO 4 INCHES 1

142 PAFZZ PAOZZ 5940-01-425-2020 30554 88-20274-1 ....TERMINAL, SPADE #6, 22-18
AWG 1

(1)	(2) SMR CODE a.	b.	(3)	(4)	(5)	(6)	(7)
ITEM NO	ARMY	AIR FORCE	NSN	CAGEC	PART NUMBER	DESCRIPTION AND USABLE ON CODE (UOC)	QTY
143	MFFZZ	MOO		30554	88-22629-036	....MARKER,IDENTIFICATION MAKE FROM P/N M23053/5-105-9 (81349), 1.25 INCHES REQUIRED	1
144	MFFZZ	MOO		30554	88-22629-037	....MARKER,IDENTIFICATION BMV 8 144 MAKE FROM P/N M23053/5-105-9 (81349), 1.25 IN. REQD	2
145	MFFZZ	MOO		30554	88-22629-038	....MARKER,IDENTIFICATION MAKE FROM P/N M23053/5-105-9 (81349), 1.25 INCHES REQUIRED	2
146	AFFFF	AOO		30554	88-22101-5	...CABLE ASSY, SHIELDED (NOT SHOWN) UOC: EMK, YNN	1
147	MFFZZ	MOO		30554	88-22101-5-1	....CABLE, SHIELDED, MAKE FROM P/N M27500-16TE2T14 (81349), 19.5 INCHES REQUIRED UOC: EMK, YNN	1
148	PAFZZ	PAOZZ	5940-01-110-6423	30554	88-20274-5	....TERMINAL, SPADE #6, 16-14 AWG UOC: EMK, YNN	2
149	PAFZZ	PAOZZ	5940-01-531-5519	30554	88-20596-1	....SHIELD, TERMINATION CUT WIRE LEAD TO 4 INCHES UOC: EMK, YNN	1
150	PAFZZ	PAOZZ	5940-01-425-2020	30554	88-20274-1	....TERMINAL, SPADE #6, 22-18 AWG UOC: EMK, YNN	1
151	MFFZZ	MOO		30554	88-22629-039	....MARKER,IDENTIFICATION MAKE FROM P/N M23053/5-105-9 (81349), 1.25 INCHES REQUIRED UOC: EMK, YNN	1
152	MFFZZ	MOO		30554	88-22629-040	....MARKER IDENTIFICATION MAKE FROM P/N M23053/5-105-9 (81349), 1.25 INCHES REQUIRED UOC: EMK, YNN	1
153	MFFZZ	MOO		30554	88-22629-041	....MARKER IDENTIFICATION MAKE FROM P/N M23053/5-105-9 (81349), 1.25 INCHES REQUIRED UOC: EMK, YNN	1
154	AFFFF	AOO		30554	88-22102-1	...CABLE ASSY, SHIELDED (NOT SHOWN)	1
155	MFFZZ	MOO		30554	88-22102-1-1	....CABLE, SHIELDED MAKE FROM P/N M27500-16TE2T14 (81349), 46.0 INCHES REQUIRED	1
156	PAFZZ	PAOZZ	5940-01-531-5519	30554	88-20596-1	....SHIELD, TERMINATION CUT WIRE LEAD TO 4 INCHES	1
157	PAFZZ	PAOZZ	5940-00-615-6073	96906	MS25036-152	....TERMINAL, LUG	1
158	PAFZZ	PAOZZ	5940-01-110-6423	30554	88-20274-5	....TERMINAL, SPADE #6, 16-14 AWG	2
159	PAFZZ	PAOZZ	5940-01-425-2020	30554	88-20274-1	....TERMINAL, SPADE #6, 22-18 AWG	1
160	MFFZZ	MOO		30554	88-22102-1-12	....INSULATION, SLEEVING BMV 8 160 MAKE FROM P/N M23053/5-105-0 (81349), 4.0 INCHES REQD	1
161	MFFZZ	MOO		30554	88-22629-053	....MARKER,IDENTIFICATION MAKE FROM P/N M23053/5-105-9 (81349), 1.25 INCHES REQUIRED	1

(1) (2) (3) (4) (5) (6) (7)

SMR CODE

a. b.

ITEM NO | ARMY AIR FORCE | NSN CAGEC PART NUMBER DESCRIPTION AND USABLE ON CODE (UOC) | QTY

162 MFFZZ MOO 30554 88-22629-054 ....MARKER,IDENTIFICATION
MAKE FROM P/N M23053/5-105-9
(81349), 1.25 INCHES REQUIRED 2

163 MFFZZ MOO 30554 88-22629-055 ....MARKER,IDENTIFICATION
MAKE FROM P/N M23053/5-105-9
(81349), 1.25 INCHES REQUIRED 2

164 AFFFF AOO 30554 88-22102-2 ...CABLE ASSY, SHIELDED (NOT
SHOWN) 1

165 MFFZZ MOO 30554 88-22102-2-1 ....CABLE, SHIELDED MAKE FROM
P/N M27500-16TE2T14 (81349), 35
INCHES REQUIRED 1

166 PAFZZ PAOZZ 5940-01-531-5519 30544 88-20596-1 ....SHIELD, TERMINATION CUT
WIRE LEAD TO 6 INCHES FOR WIRE
TERMINATING AT A5-GP. CUT WIRE
LEAD TO 7 INCHES 2

167 PAFZZ PAOZZ 5940-01-110-6423 30554 88-20274-5 ....TERMINAL, SPADE #6, 16-14
AWG 2

168 PAFZA PAOZA 5999-01-463-4017 30554 88-21943 ....PIN 2

169 PAFZZ PAOZZ 5999-01-366-5119 30554 88-21941 ....PIN .... 1 170 PAFZZ PAOZZ 5940-01-369-2270
30554 88-20274-4 ....TERMINAL, SPADE #10, 22-18
AWG 1

171 MFFZZ MOO 30554 88-22629-056 ....MARKER,IDENTIFICATION
MAKE FROM P/N M23053/5-105-9
(81349), 1.25 INCHES REQUIRED 1

172 MFFZZ MOO 30554 88-22629-057 ....MARKER,IDENTIFICATION
MAKE FROM P/N M23053/5-105-9
(81349), 1.25 INCHES REQUIRED 2

173 MFFZZ MOO 30554 88-22629-058 ....MARKER,IDENTIFICATION
MAKE FROM P/N M23053/5-105-9
(81349), 1.25 INCHES REQUIRED 2

174 AFFFF AOO 30554 88-22102-3 ...CABLE ASSY, SHIELDED (NOT
SHOWN) 1

175 MFFZZ MOO 30554 88-22102-3-1 ....CABLE, SHIELDED MAKE FROM
P/N M27500-16TE2T14 (81349), 23
INCHES REQUIRED 1

176 PAFZZ PAOZZ 5940-01-531-5519 30554 88-20596-1 ....SHIELD, TERMINATION CUT
WIRE LEAD TO 4 INCHES FOR WIRE
TERMINATING AT TB4-9. CUT WIRE
LEAD TO 7 INCHES 2

177 PAFZZ PAOZZ 5940-01-110-6423 30554 88-20274-5 ....TERMINAL, SPADE #6, 16-14
AWG 2

178 PAFZZ PAOZZ 5940-01-425-2020 30554 88-20274-1 ....TERMINAL, SPADE #6, 22-18
AWG 1

179 PAFZA PAOZA 5999-01-463-4017 30554 88-21943 ....PIN 2

180 PAFZZ PAOZZ 5999-01-366-5119 30554 88-21941 ....PIN .... 1 181 MFFZZ MOO 30554 88-22629-059
....MARKER,IDENTIFICATION
MAKE FROM P/N M23053/5-105-9
(81349), 1.25 INCHES REQUIRED 1

182 MFFZZ MOO 30554 88-22629-060 ....MARKER,IDENTIFICATION
MAKE FROM P/N M23053/5-105-9
(81349), 1.25 INCHES REQUIRED 2

183 MFFZZ MOO 30554 88-22629-061 ....MARKER,IDENTIFICATION
MAKE FROM P/N M23053/5-105-9
(81349), 1.25 INCHES REQUIRED 2

184 AFFFF AOO 30554 88-22104 ...CABLE ASSY, SHIELDED (NOT
SHOWN) 1

(1) ITEM NO	(2) SMR CODE a. ARMY	b. AIR FORCE	(3) NSN	(4) CAGEC	(5) PART NUMBER	(6) DESCRIPTION AND USABLE ON CODE (UOC)	(7) QTY
185	MFFZZ	MOO		30554	88-22104-1	....CABLE, SHIELDED MAKE FROM P/N M27500-20TE3T14 (81349), 53 INCHES REQUIRED	1
186	PAFZZ	PAOZZ	5940-01-425-2020	30554	88-20274-1	....TERMINAL, SPADE #6, 22-18 AWG	3
187	PAFZZ	PAOZZ	5940-01-136-0953	30554	88-20596-3	....SHIELD, TERMINATION CUT WIRE LEAD TO 7 INCHES	1
188	PAFZZ	PAOZZ	5940-01-369-2270	30554	88-20274-4	....TERMINAL, SPADE #10, 22-18 AWG	1
189	MFFZZ	MOO		30554	88-22629-077	....MARKER,IDENTIFICATION MAKE FROM P/N M23053/5-105-9 (81349), 1.25 INCHES REQUIRED	1
190	MFFZZ	MOO		30554	88-22629-078	....MARKER IDENTIFICATION MAKE FROM P/N M23053/5-105-9 (81349), 1.25 INCHES REQUIRED	1
191	MFFZZ	MOO		30554	88-22629-079	....MARKER IDENTIFICATION MAKE FROM P/N M23053/5-105-9 (81349), 1.25 INCHES REQUIRED	1
192	MFFZZ	MOO		30554	88-22629-080	....MARKER IDENTIFICATION MAKE FROM P/N M23053/5-105-9 (81349), 1.25 INCHES REQUIRED	1
193	AFFFF	AOO		30554	88-22630	...CABLE ASSY, SHIELDED (NOT SHOWN)	1
194	MFFZZ	MOO		30554	88-22630-1	....CABLE, SHIELDED MAKE FROM P/N M27500-16TE2T14 (81349), 24.5 INCHES REQUIRED	1
195	MFFZZ	MOO		30554	88-22630-2	....CABLE, SHIELDED MAKE FROM P/N M27500-16TE1T14 (81349), 23.5 INCHES REQUIRED	1
196	PAFZA	PAOZA	5999-01-463-4017	30554	88-21943	....PIN	3
197	PAFZZ	PAOZZ	5940-01-135-7081	30554	88-20596-2	....SHIELD, TERMINATION CUT WIRE LEAD TO 4 INCHES	1
198	PAFZZ	PAOZZ	5940-01-531-5519	30554	88-20596-1	....SHIELD, TERMINATION CUT WIRE LEAD TO 7 INCHES FOR WIRE TERMINATING AT J6-9. CUT WIRE LEAD TO 4	2
199	PAFZZ	PAOZZ	5940-01-110-6423	30554	88-20274-5	....TERMINAL, SPADE #6, 16-14 AWG	3
200	PAFZZ	PAOZZ	5940-01-425-2020	30554	88-20274-1	....TERMINAL, SPADE #6, 22-18 AWG	1
201	PAFZZ	PAOZZ	5999-01-366-5119	30554	88-21941	....PIN	1
202	MFFZZ	MOO		30554	88-22629-085	....MARKER,IDENTIFICATION MAKE FROM P/N M23053/5-105-9 (81349), 1.25 INCHES REQUIRED	2
203	MFFZZ	MOO		30554	88-22629-082	....MARKER,IDENTIFICATION MAKE FROM P/N M23053/5-105-9 (81349), 1.25 INCHES REQUIRED	2
204	MFFZZ	MOO		30554	88-22629-086	....MARKER,IDENTIFICATION MAKE FROM P/N M23053/5-105-9 (81349), 1.25 INCHES REQUIRED	2
205	MFFZZ	MOO		30554	88-22629-084	....MARKER,IDENTIFICATION MAKE FROM P/N M23053/5-105-9 (81349), 1.25 INCHES REQUIRED	2
206	PAFZZ	PAOZZ	5999-01-092-2655	30554	88-20476	...CONTACT,ELECTRICAL UOC: 86Q, 86R	3
207	PAFZZ	PAOZZ		30554	88-20473	...CONNECTOR,BODY UOC: 86Q, 86R	1

(1) ITEM NO	(2) SMR CODE a. ARMY	b. AIR FORCE	(3) NSN	(4) CAGEC	(5) PART NUMBER	(6) DESCRIPTION AND USABLE ON CODE (UOC)	(7) QTY
208	AFFFF	AOO		30554	97-24133	...CABLE ASSY, SHIELDED (NOT SHOWN) UOC: 86Q, 86R	1
209	MFFZZ	MOO		30554	88-20574-2	....CABLE, SHIELDED,19.5INCHES REQUIRED UOC: 86Q, 86R	1
210	PAFZZ	PAOZZ	5940-01-110-6423	30554	88-20274-5	....TERMINAL, SPADE #6, 16-14 AWG UOC: 86Q, 86R	2
211	PAFZZ	PAOZZ	5940-01-531-5519	30554	88-20596-1	....SHIELD, TERMINATION CUT WIRE LEAD TO 4 INCHES UOC: 86Q, 86R	1
212	PAFZZ	PAOZZ	5940-01-425-2020	30554	88-20274-1	....TERMINAL, SPADE #6, 22-18 AWG UOC: 86Q, 86R	1
213	MFFZZ	MOO		30554	88-22629-087	....MARKER,INDENTIFICATION MAKE FROM P/N M23053/5-105-9 (81349), 1.25 INCHES REQUIRED UOC: 86Q, 86R	1
214	MFFZZ	MOO		30554	88-22629-040	....MARKER,INDENTIFICATION MAKE FROM P/N M23053/5-105-9 (81349), 1.25 INCHES REQUIRED UOC: 86Q, 86R	1
215	MFFZZ	MOO		30554	88-22629-041	....MARKER,INDENTIFICATION MAKE FROM P/N M23053/5-105-9 (81349), 1.25 INCHES REQUIRED UOC: 86Q, 86R	1
216	PAFZZ	PAOZZ	5940-01-259-2190	30554	88-20274-11	....TERMINAL, SPADE #8, 16-14 AWG UOC: 86Q, 86R	1

END OF FIGURE

FIELD AND SUSTAINMENT MAINTENANCE

15 kW 50/60 AND 400 Hz SKID MOUNTED TACTICAL QUIET GENERATOR SETS
GROUP 03 CONTROL BOX ASSEMBLY: CONTROL BOX PANELS

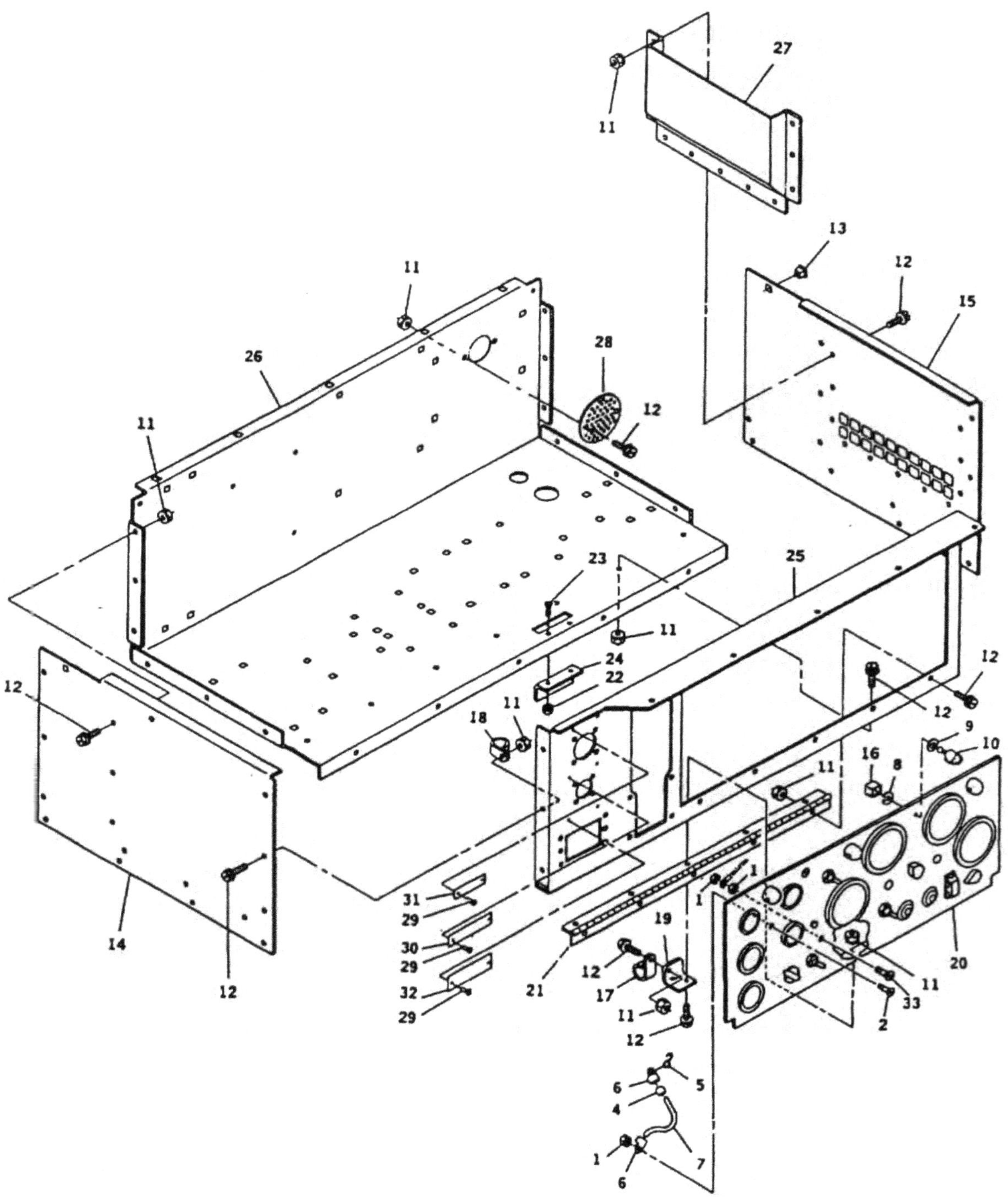

Figure 9. Control Box Panels.

(1) ITEM NO	(2) SMR CODE a. ARMY b. AIR FORCE	(3) NSN	(4) CAGEC	(5) PART NUMBER	(6) DESCRIPTION AND USABLE ON CODE (UOC)	(7) QTY

GROUP 03 CONTROL BOX ASSEMBLY

FIG. 9 CONTROL BOX PANELS

1 PAFZZ PAOZZ 5310-00-052-3632 30554 69-561-3 . . NUT, PLAIN, ASSEMBLED 1
2 PAFZZ PAOZZ 5305-00-954-5629 80205 MS35198-42 . . SCREW, MACHINE .... 1 3 AFFFF AOO 30554 88-22120 . . HOLDER, CONTROL PAN (NOT
SHOWN) 1
4 MFFZZ MOO 30554 88-22120-4 ...INSULATION, SLEEVING MAKE
FROM P/N M23053/5-104-0 (81349), 1.5 INCHES REQUIRED 1
5 PAFZZ PAOZZ 4030-00-780-9350 96906 MS87006-13 ...HOOK,CHAIN,S 1
6 PAFZZ PAOZZ 5940-00-115-2677 96906 MS20659-144 ...TERMINAL, LUG .... 2 7 MFFZZ MOO 30554 88-22120-2 ...ROPE, FIBROUS BMV 97MAKE
FROM P/N 1-41NDIA (72205), 10 INCHES REQD 1
8 PAFZZ PAOZZ 5310-00-822-8525 30554 88-20052 . . RETAINER 2
9 PAFZZ PAOZZ 5310-01-365-4381 30554 88-20242 . . WASHER, WEAR 2
10 PAFZZ PAOZZ 5325-01-161-2654 30554 88-20051 . . STUD, ¼ TURN 2
11 PAFZZ PAOZZ 5310-01-012-3595 30554 69-561-6 . . NUT, PLAIN, ASSEMBLED 36
12 PAFZZ PAOZZ 5306-01-156-7663 30554 88-20260-21 . SCREW, HEX WASHERHEAD 36
13 PAFZZ PAOZZ 5310-00-903-8595 30554 88-21674-2 . . NUT, CAGE .... 7 14 XDFZZ XB 30554 88-21868 . . PANEL, CONTROL BOX LEFT
SIDE 1
15 XDFZZ XB 30554 88-21869 . . PANEL, CONTROL BOX RIGHT
SIDE 1
16 PAFZZ PAOZZ 5325-01-301-7903 94222 85-35-309-56 . . RECEPTACLE, TURNLOCK 2
17 PAFZZ PAOZZ 5340-00-724-7038 96906 MS21333-76 . . CLAMP, LOOP 1
18 PAFZZ PAOZZ 5340-00-050-2740 96906 MS21333-75 . . CLAMP, LOOP 1
19 XDFZZ XB 30554 88-21980 . . BRACKET, HARNESS SU 1
20 XDFZZ XB 30554 88-22042 . . PANEL, CONTROL .... 1 21 MFFZZ MOO 30554 88-21876 . . HINGE, CONTROL PANEL MAKE
FROM P/N Z60960 (03007), 21.75 INCHES REQUIRED 1
22 PAFZZ PAOZZ 5310-00-063-7360 30554 69-561-2 . . NUT, PLAIN, ASSEMBLED 2
23 PAFZZ PAOZZ 5305-01-467-1561 30554 88-22791-2 . . SCREW, MACHINE 2
24 XDFZZ XB 30554 88-21874 . . ANGLE, LATCH 1
25 XDFZZ XB 30554 88-21873 . . FRAME, CONTROL PANEL 1
26 XDFZZ XB 30554 88-21871 . . BOTTOM, CONTROL BOX 1
27 XDFZZ XB 30554 88-22772 . . BAFFLE, CONTROL BOX 1
28 XDFZZ XB 30554 88-22774 . . SCREEN, VENT 1
29 PAFZZ PAOZZ 5320-01-019-5694 81349 M24243/1A402 . . RIVET, BLIND .... 6 30 MDFZZ MDO 30554 88-20074 . . PLATE, IDENTIFICATION, PARAL-
LELING RECEPTACLE 1
31 MDFZZ MDO 9905-01-372-7986 30554 88-20197 . . PLATE, IDENTIFICATION, DIA-
GNOSTICS RECEPTACLE 1
32 MDFZZ MDO 30554 88-20073 . . PLATE IDENTIFICATION, CONVENI-
ENCE RECEPTACLE
UOC: EMK, 86Q............................ 1
32 MDFZZ MDO 30554 88-22737 . . PLATE, IDENTIFICATION, CON-
VENIENCE RECEPTACLE
UOC: YNN, 86R............................ 1

(1) ITEM NO	(2) SMR CODE a. ARMY	b. AIR FORCE	(3) NSN	(4) CAGEC	(5) PART NUMBER	(6) DESCRIPTION AND USABLE ON CODE (UOC)	(7) QTY
33	PAFZZ	PAOZZ	5305-00-954-5638	80205	MS35198-46	. . SCREW, MACHINE	1

END OF FIGURE

FIELD AND SUSTAINMENT MAINTENANCE

15 kW 50/60 AND 400 Hz SKID MOUNTED TACTICAL QUIET GENERATOR SETS

GROUP 04 AIR INTAKE AND EXHAUST SYSTEM: MUFFLER AND PIPES

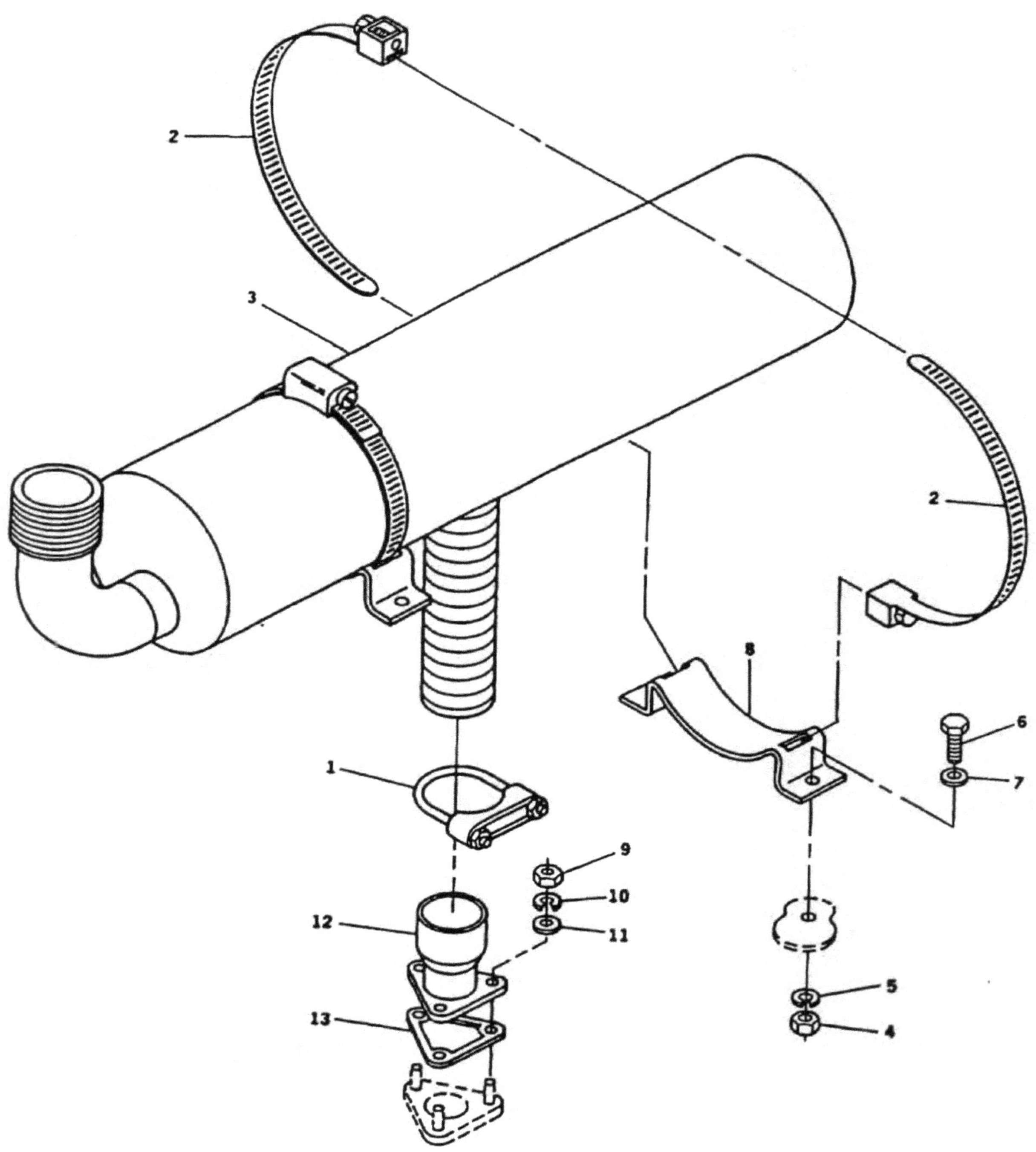

Figure 10. Muffler and Pipes.

(1) ITEM NO	(2) SMR CODE a. ARMY	b. AIR FORCE	(3) NSN	(4) CAGEC	(5) PART NUMBER	(6) DESCRIPTION AND USABLE ON CODE (UOC)	(7) QTY
						GROUP 04 AIR INTAKE AND EXHAUST SYSTEM	
						FIG. 10 MUFFLER AND PIPES	
1	PAFZZ	PAOZZ	5340-01-367-1503	30554	88-21975	. CLAMP, LOOP	1
2	PAFZZ	PAOZZ	4730-01-470-1578	30554	88-20561-6	. CLAMP, HOSE	4
3	PAFZZ	PAOZZ	2990-01-369-2079	30554	88-21959	. MUFFLER ASSY, EXHAUST	1
4	PAFZZ	PAOZZ	5310-01-466-6312	30554	88-22790-1	. NUT, PLAIN, HEX	4
5	PAFZZ	PAOZZ	5310-00-274-8715	80205	MS35338-63	. WASHER, LOCK	4
6	PAFZZ	PAOZZ	5305-01-531-4346	30554	88-20260-31	. SCREW, HEX WASHERHEAD	4
7	PAFZZ	PAOZZ	5310-01-103-6042	96906	MS51412-4	. WASHER, FLAT	4
8	XDFZZ	XB	2990-01-390-4457	30554	88-21712	. HANGER, ENGINE EXHAUST	2
9	PAFZZ	PAOZZ	5310-01-366-3539	80204	B18241B100	. NUT, PLAIN, HEX UOC: EMK, YNN	3
10	PAFZZ	PAOZZ	5310-01-368-3048	30554	88-22331-2	. WASHER, LOCK UOC: EMK, YNN	3
11	PAFZZ	PAOZZ	5310-01-257-7590	96906	MS51412-7	. WASHER, FLAT UOC: EMK, YNN	3
12	PAFZZ	PAOZZ	2990-01-374-9149	30554	88-22183	. CONNECTOR, EXHAUST UOC: EMK, YNN	1
13	PAFZZ	PAOZZ	5330-01-368-5932	30554	88-22182	. GASKET UOC: EMK, YNN	1

END OF FIGURE

FIELD AND SUSTAINMENT MAINTENANCE

15 kW 50/60 AND 400 Hz SKID MOUNTED TACTICAL QUIET GENERATOR SETS
GROUP 04 AIR INTAKE AND EXHAUST SYSTEM: TURBOCHARGER INSTALLATION

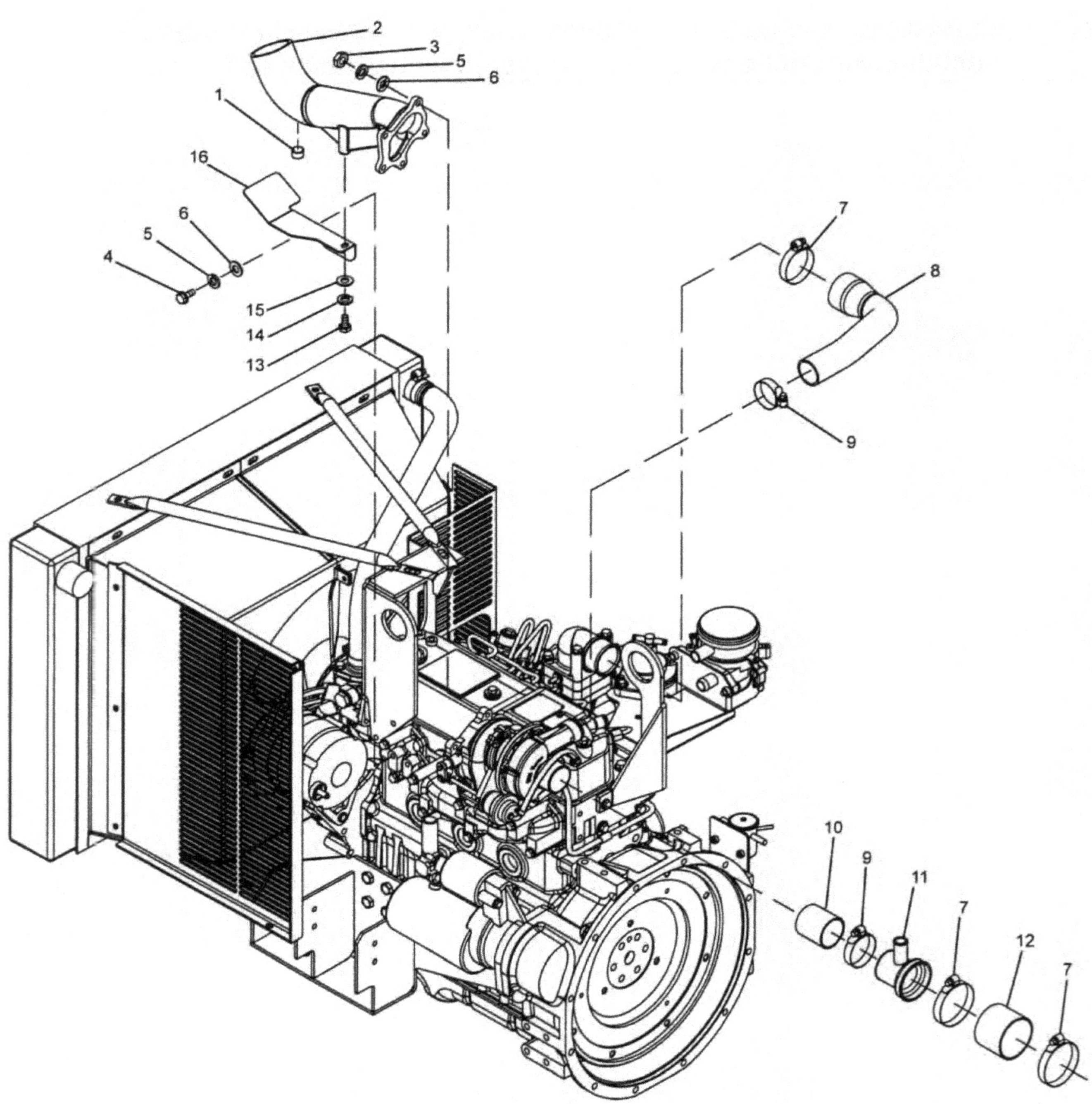

Figure 11. Turbocharger Installation.

(1) ITEM NO	(2) SMR CODE a. ARMY	(2) SMR CODE b. AIR FORCE	(3) NSN	(4) CAGEC	(5) PART NUMBER	(6) DESCRIPTION AND USABLE ON CODE (UOC)	(7) QTY
						GROUP 04 AIR INTAKE AND EXHAUST SYSTEM	
						FIG. 11 TURBOCHARGER INSTALLATION	
1	PAFZZ	PAFZZ	4730-00-089-2515	81343	4-130109E-C	. PLUG, PIPE, HEX UOC: 86Q, 86R	1
2	XBFZZ	XBFZZ		30554	97-24021	. MANIFOLD, EXHAUST UOC: 86Q, 86R	1
3	PAFZZ	PAFZZ		30554	97-24111-9	. NUT, HEXAGON UOC: 86Q, 86R	5
4	PAFZZ	PAFZZ	5306-01-179-1438	80204	B18231A1	. BOLT, MACHINE UOC: 86Q, 86R	2
5	PAFZZ	PAFZZ	5310-01-099-9532	30554	88-22331-1	. WASHER, LOCK UOC: 86Q, 86R	7
6	PAFZZ	PAFZZ	5310-01-531-4937	30554	88-20033-24A	. WASHER, FLAT UOC: 86Q, 86R	7
7	PAFZZ	PAFZZ	4730-01-470-1595	30554	88-20561-4	. CLAMP, HOSE UOC: 86Q, 86R	3
8	PAFZZ	PAFZZ		30554	97-24022	. HOSE, TURBOCHARGER UOC: 86Q, 86R	1
9	PAFZZ	PAFZZ	4730-01-470-1567	30554	88-20561-3	. CLAMP, HOSE UOC: 86Q, 86R	2
10	PAFZZ	PAFZZ		30554	97-24029	. HOSE, 1.75 X 2.00 UOC: 86Q, 86R	1
11	PAFZZ	PAFZZ		30554	97-24028	. REDUCER, AIR CLEANER UOC: 86Q, 86R	1
12	PAFZZ	PAFZZ		30554	97-24030	. HOSE, 2.50 X 2.00 UOC: 86Q, 86R	1
13	PAFZZ	PAFZZ	5305-01-369-2166	30554	88-20260-30	. BOLT, MACHINE UOC: 86Q, 86R	1
14	PAFZZ	PAFZZ		30554	88-20556-38	. WASHER, LOCK UOC: 86Q, 86R	1
15	PAFZZ	PAFZZ		30554	88-20033-21A	. WASHER, FLAT UOC: 86Q, 86R	1
16	PAFZZ	PAFZZ		30554	97-24020	. BRACKET, MANIFOLD UOC: 86Q, 86R	1
						END OF FIGURE	

FIELD AND SUSTAINMENT MAINTENANCE

15 kW 50/60 AND 400 Hz SKID MOUNTED TACTICAL QUIET GENERATOR SETS

GROUP 04 AIR INTAKE AND EXHAUST SYSTEM: AIR CLEANER AND PIPING

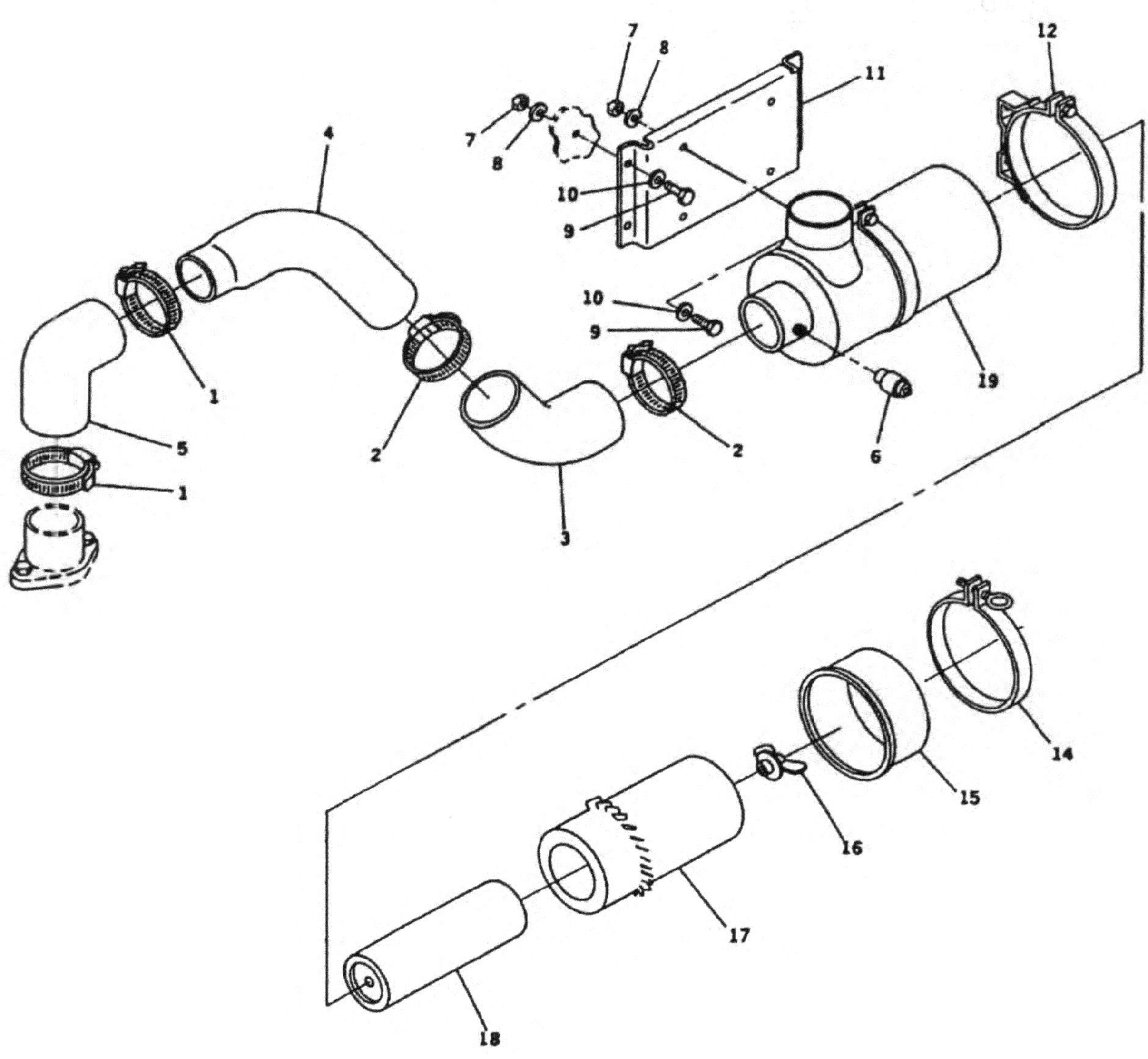

Figure 12. Air Cleaner and Piping.

(1)	(2) SMR CODE a.	b.	(3)	(4)	(5)	(6)	(7)
ITEM NO	ARMY	AIR FORCE	NSN	CAGEC	PART NUMBER	DESCRIPTION AND USABLE ON CODE (UOC)	QTY

GROUP 04 AIR INTAKE AND EXHAUST SYSTEM

FIG. 12 AIR CLEANER AND PIPING

1 PAFZZ PAOZZ 4730-01-470-1595 30554 88-20561-4 . CLAMP, HOSE 2
2 PAFZZ PAOZZ 4730-01-470-1423 30554 88-20561-5 . CLAMP, HOSE 2
3 PAFZZ PAOZZ 4720-01-375-1929 30554 88-21177-1 . HOSE, PREFORMED 1
4 XDFZZ XB 4710-01-392-8846 30554 88-22301 . TUBE ASSY, AIR INTAKE 1
5 PAFZZ PAOZZ 4720-01-366-6257 30554 88-22300 . HOSE, PREFORMED 1
6 PAFZZ PAOZZ 2940-01-531-4912 30554 88-21127 . INDICATOR ASSY, AIR CL 1
7 PAFZZ PAOZZ 5310-01-466-6312 30554 88-22790-1 . NUT, PLAIN, HEX 8
8 PAFZZ PAOZZ 5310-00-274-8715 80205 MS35338-63 . WASHER, LOCK 8
9 PAFZZ PAOZZ 5305-01-056-1501 30554 88-20260-32 . SCREW, HEX WASHERHEAD 8
10 PAFZZ PAOZZ 5310-01-103-6042 96906 MS51412-4 . WASHER, FLAT .... 8 11 XDFZZ XB 30554 88-21743 . SUPPORT, AIR CLEANER
UOC: EMK, YNN 1
11 XDFZZ XB 30554 97-24042 . SUPPORT, AIR CLEANER
UOC: 86Q, 86R............................. 1
12 PAFZZ PAOZZ 5340-01-074-8126 30554 88-21129 . BAND, AIR CLEANER MNTG. 2
13 XDFFF XB 30554 88-21065 . AIR CLEANER (NOT SHOWN) 1
14 PAFZZ PAOZZ 5340-01-071-8006 18265 P00-2904 . . CLAMP ASSY 1
15 XDFZZ XB 18265 P12-0316 . . CUP ASSY 1
16 PAFZZ PAOZZ 5310-01-190-4614 18265 P10-1870 . . NUT ASSY 2
17 PAFZZ PAOZZ 2940-01-103-3268 18265 SMP18-1072 . . FILTER ELEMENT, INTA 1
18 PAFZZ PAOZZ 2940-01-103-3267 18265 P12-0307 . . FILTER ELEMENT, INTA 1
19 XAFZZ XA 18265 P12-028 . . CANISTER, AIR CLEAN 1

END OF FIGURE

FIELD AND SUSTAINMENT MAINTENANCE

15 kW 50/60 AND 400 Hz SKID MOUNTED TACTICAL QUIET GENERATOR SETS

GROUP 05 COOLANT SYSTEM

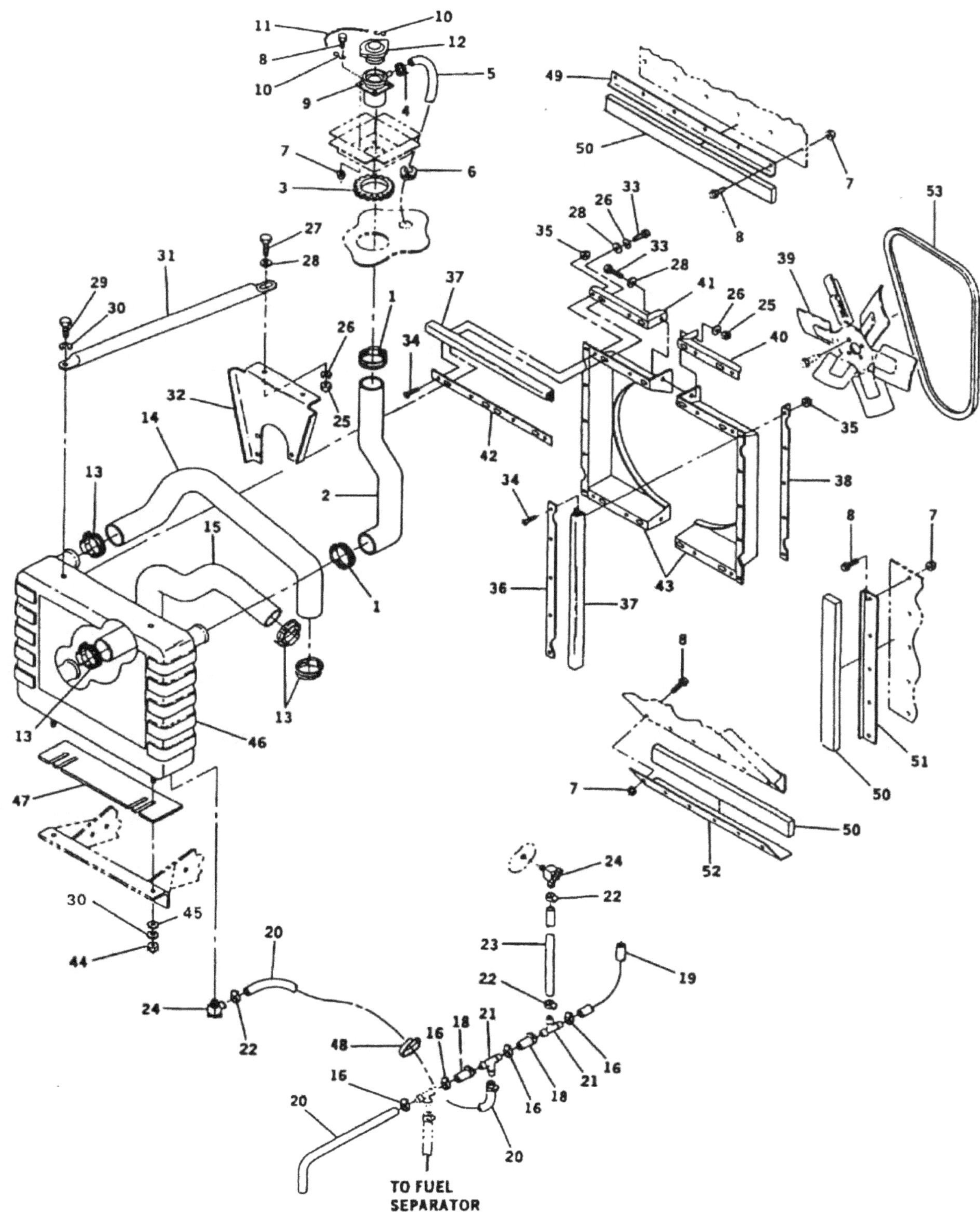

Figure 13. Coolant System (Sheet 1 of 2).

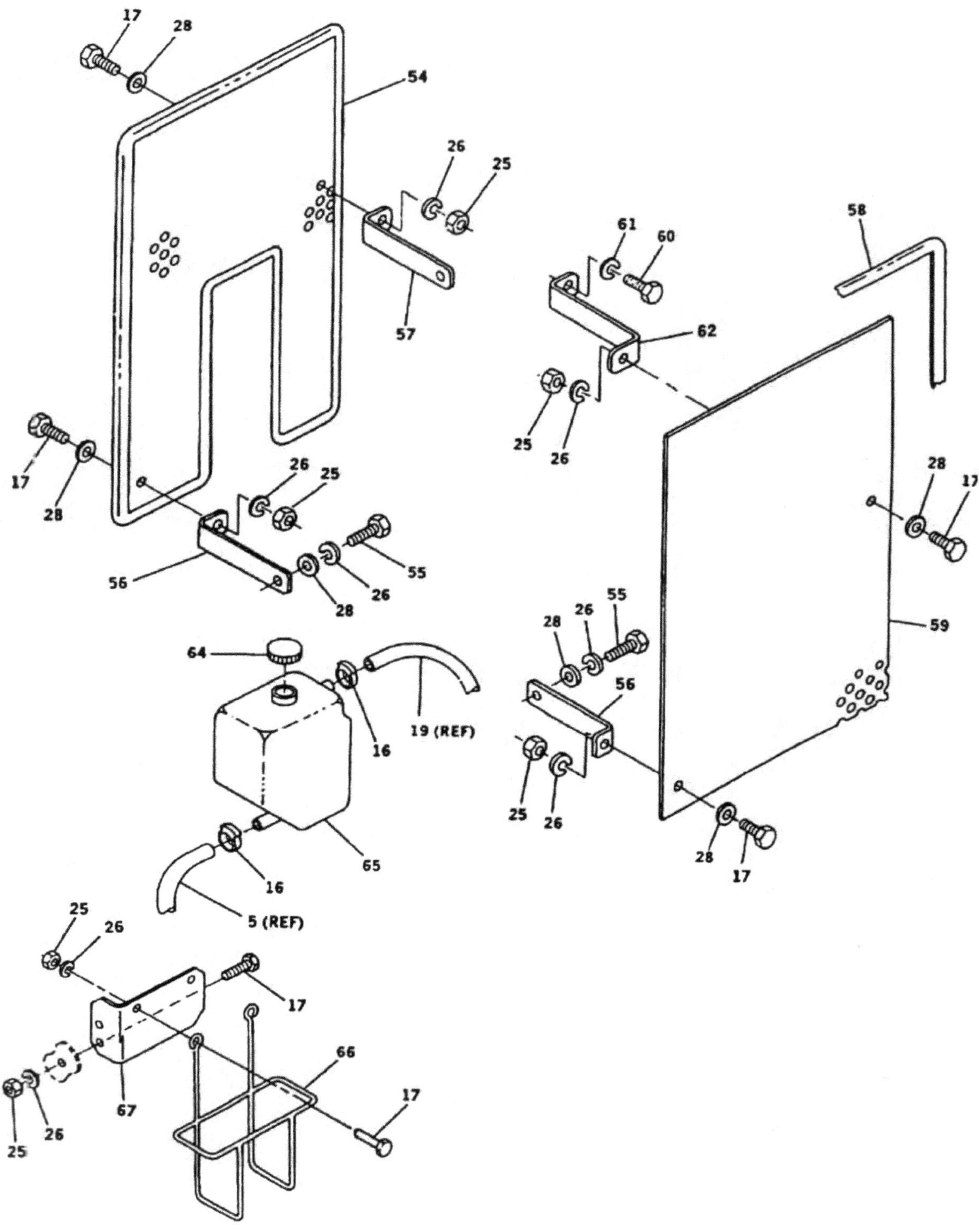

Figure 13. Coolant System (Sheet 2 of 2).

(1) (2) (3) (4) (5) (6) (7)
SMR CODE
a. b.
ITEM NO ARMY AIR FORCE NSN CAGEC PART NUMBER DESCRIPTION AND USABLE ON CODE (UOC) QTY

GROUP 05 COOLANT SYSTEM

FIG. 13 COOLANT SYSTEM

1 PAFZZ PAOZZ 4730-01-470-1595 30554 88-20561-4 . CLAMP, HOSE 2
2 PAFZZ PAOZZ 4720-01-374-0783 30554 88-22170 . HOSE, PREFORMED .... 1 3 MFFZZ MOO 30554 88-21604-228 . GROMMET, NONMETALLIC MAKE
FROM P/N 88-20543-2 (30554), AS REQUIRED
UOC: EMK 1
3 MFFZZ MOO 30554 97-24604-228 . GROMMET, NONMETALLIC MAKE
FROM P/N 88-20543-2 (30554), AS REQUIRED
UOC: 86Q................................. 1
3 MFFZZ MOO 30554 88-21605-228 . GROMMET, NONMETALLIC MAKE
FROM P/N 88-20543-2 (30554), AS REQUIRED
UOC: YNN 1
3 MFFZZ MOO 30554 97-24605-228 . GROMMET, NONMETALLIC MAKE
FROM P/N 88-20543-2 (30554), AS REQUIRED
UOC: 86R 1
4 PAFZZ PAOZZ 4730-00-908-3195 81343 SAE J1508-06 . CLAMP, HOSE .... 1 5 MFFZZ MOO 30554 88-22302-12 . HOSE, NONMETALLIC MAKE FROM
P/N MS521302A101360 (96906), 53 INCHES REQUIRED. RADIATOR OVERFLOW 1
6 PAFZZ PAOZZ 5325-00-174-9038 96906 MS35489-20 . GROMMET, NONMETALLIC 1
7 PAFZZ PAOZZ 5310-01-012-3595 30554 69-561-6 . NUT, PLAIN, ASSEMBLED 4
8 PAFZZ PAOZZ 5306-01-156-7663 30554 88-20260-21 . SCREW, HEX WASHERHEAD 4
9 XDFZZ XB 30554 88-21849 . NECK, FILL RADIATOR 1
10 PAFZZ PAOZZ 4030-00-236-0843 96906 MS87006-11 . HOOK, CHAIN, S .... 2 11 MFFZZ MOO 30554 88-21604-272 . CHAIN, CAP MAKE FROM P/N
RRC271TY2CL7-16 (81348), 6.5 INCHES REQD
UOC: EMK 1
11 MFFZZ MOO 30554 97-24604-272 . CHAIN, CAP MAKE FROM P/N
RRC271TY2CL7-16 (81348), 6.5 INCHES REQD
UOC: 86Q................................. 1
11 MFFZZ MOO 30554 88-21605-272 . CHAIN, CAP MAKE FROM P/N
RRC271TY2CL7-16 (81348), 6.5 INCHES REQD
UOC: YNN 1
11 MFFZZ MOO 30554 97-24605-272 . CHAIN, CAP MAKE FROM P/N
RRC271TY2CL7-16 (81348), 6.5 INCHES REQD
UOC: 86R 1
12 XCFZZ XCOZZ 5340-01-368-6063 30554 88-22533 . CAP, FILLER OPENING ORDER NSN 5340-01-340-8706, P/N AX-2950 (78225), AND SOLDER LUG P/N 5413-28 (86928), TM 9-6115-643-24 1
13 PAFZZ PAOZZ 4730-01-470-1567 30554 88-20561-3 . CLAMP, HOSE
UOC: EMK, YNN 4

(1)	(2) SMR CODE		(3)	(4)	(5)	(6)	(7)
ITEM NO	a. ARMY	b. AIR FORCE	NSN	CAGEC	PART NUMBER	DESCRIPTION AND USABLE ON CODE (UOC)	QTY

14 PAFZZ PAOZZ 4720-01-385-1102 30554 88-22171 . HOSE, PREFORMED RADIATOR UPPER UOC: EMK, YNN 1

15 PAFZZ PAOZZ 4720-01-368-5430 30554 88-22172 . HOSE, PREFORMED RADIATOR LOWER 1

16 PAFZZ PAOZZ 4730-01-370-5426 30554 88-22554-2 . CLAMP, HOSE UOC: EMK, YNN 8

17 PAFZZ PAOZZ 5305-00-068-0508 80204 B1821BH025C075N . SCREW, CAP, HEX HEAD .... 10 18 MFFZZ MOO 30554 88-22302-4 . HOSE, NONMETALLIC MAKE FROM P/N MS521302A101360 (96906), 31 INCHES REQUIRED 2

19 MFFZZ MOO 30554 88-22302-6 . HOSE, NONMETALLIC MAKE FROM P/N MS521302A101360 (96906), 20 INCHES REQUIRED 2

20 MFFZZ MOO 30554 88-22752-3 . HOSE, NONMETALLIC MAKE FROM P/N 483666 (49185), 12 INCHES REQUIRED 1

21 PAFZZ PAOZZ 4730-01-531-4884 30554 88-22323 . TEE, HOSE 2

22 PAFZZ PAOZZ 4730-01-378-5224 30554 88-22554-3 . CLAMP, HOSE .... 4 23 MFFZZ MOO 30554 88-22752-2 . HOSE, COOLANT DRAIN MAKE FROM P/N 483666 (49185), 20 INCHES REQUIRED 1

24 PAFZZ PAOZZ 4820-01-381-5079 30554 88-22751 . COCK, DRAIN UOC: EMK, YNN 2

25 PAFZZ PAOZZ 5310-01-466-6312 30554 88-22790-1 . NUT, PLAIN, HEX 18

26 PAFZZ PAOZZ 5310-00-274-8715 80205 MS35338-63 . WASHER, LOCK 24

27 PAFZZ PAOZZ 5305-01-056-1501 30554 88-20260-32 . SCREW, HEX WASHERHEAD 2

28 PAFZZ PAOZZ 5310-01-103-6042 96906 MS51412-4 . WASHER, FLAT 26

29 PAFZZ PAOZZ 5305-01-458-1624 80204 B1821BH038C062N . SCREW, CAP, HEX HEAD 2

30 PAFZZ PAOZZ 5310-00-011-5093 80205 MS35338-65 . WASHER, LOCK 4

31 XDFZZ XB 30554 88-21700 . TIE ROD, RADIATOR U .... 2 32 XDFZZ XB 30554 88-22424 . SUPPORT, RADIATOR UOC: EMK, YNN 1

33 PAFZZ PAOZZ 5305-00-068-0509 80204 B1821BH025C125N . SCREW, CAP, HEX HEAD UOC: EMK, YNN 6

34 PAFZZ PAOZZ 5305-01-466-4406 96906 MS51481-06 . SCREW, MACHINE 18

35 PAFZZ PAOZZ 5310-00-811-3494 80205 MS21044N08 . NUT, SELF LOCKING .... 18 36 XDFZZ XB 30554 88-22727 . SUPPORT, SIDE RADIATOR UOC: EMK, YNN 2

37 MFFZZ MOO 30554 88-21604-364 . SEAL, RADIATOR MAKE FROM P/N 88-22712 (30554), AS REQUIRED UOC: EMK 1

37 MFFZZ MOO 30554 88-21605-364 . SEAL, RADIATOR MAKE FROM P/N 88-22712 (30554), AS REQUIRED UOC: YNN 1

38 XDFZZ XB 30554 88-22729 . STIFFENER, SIDE RADIATOR UOC: EMK, YNN 2

39 XDFZZ XB 2930-01-466-8474 30554 88-22668 . FAN, ENGINE COOLING .... 1 40 XDFZZ XB 30554 88-22725 . STIFFENER, RADIATOR LEFT SIDE UOC: EMK, YNN 2

41 XDFZZ XB 30554 88-22726 . STIFFENER, RADIATOR RIGHT SIDE UOC: EMK, YNN 2

42 XDFZZ XB 30554 88-22728 . SUPPORT, RADIATOR SE UOC: EMK, YNN 2

(1) (2) (3) (4) (5) (6) (7)
SMR CODE
a. b.
ITEM NO ARMY AIR FORCE NSN CAGEC PART NUMBER DESCRIPTION AND USABLE ON CODE (UOC) QTY

43 XDFZZ XB 4140-01-466-8879 30554 88-21695 . SHROUD, FAN
UOC: EMK, YNN 2
44 PAFZZ PAOZZ 5310-01-466-6687 30554 88-22790-3 . NUT, PLAIN, HEX
UOC: EMK, YNN 2
45 PAFZZ PAOZZ 5310-01-257-7590 96906 MS51412-7 . WASHER, FLAT 2
46 PAFFF PAOOO 2930-01-368-1071 30554 88-22167 . RADIATOR, ENGINE .... 1 47 XDFZZ XB 5365-01-381-3773 30554 88-22703 . SPACER, SPECIAL
UOC: EMK, YNN 1
48 PAFZZ PAOZZ 5975-00-074-2072 96906 MS3367-1-9 . STRAP, TIEDOWN, ELECT
UOC: EMK, YNN 1
49 XDFZZ XB 30554 88-22723 . SUPPORT, UPPER FRONT
UOC: EMK, YNN 1
50 MFFZZ MOO 30554 88-21604-317 . SEAL, DOOR MAKE FROM P/N
6130TZS-S1SXPSA1S .75X1.00
(99739), AS REQUIRED
UOC: EMK 1
50 MFFZZ MOO 30554 88-21605-317 . SEAL, DOOR MAKE FROM P/N
6130TZS-S1SXPSA1S .75X1.00
(99739), AS REQUIRED
UOC: YNN 1
51 XDFZZ XB 30554 88-22724 . SUPPORT, SIDE FRONT
UOC: EMK, YNN 2
52 XDFZZ XB 30554 88-22722 . SUPPORT, LOWER RADIA
UOC: EMK, YNN 1
53 PAFZZ PAOZZ 3030-01-463-9774 30554 88-22740 . BELT, V
UOC: EMK, YNN 1
54 XDFZZ XB 4140-01-507-4176 30554 88-22490 . FAN, GUARD RIGHT SIDE
UOC: EMK, YNN 1
55 PAFZZ PAOZZ 5305-00-225-3843 80204 B1821BH025C100N . SCREW, CAP, HEX HEAD .... 4 56 XDFZZ XB 30554 88-22492 . FAN, GUARD SUPPORT
UOC: EMK, YNN 4
57 XDFZZ XB 30554 88-22491 . BRACKET .... 1 58 MFFZZ MOO 9390-01-470-1205 30554 88-20187 . PROTECTOR, EDGE BMV12 58
MAKE FROM P/N 62-B7-1/16 (57137),
AS REQUIRED
UOC: EMK, YNN 1
59 XDFZZ XB 4140-01-507-4174 30554 88-22489 . FAN GUARD LS LEFT SIDE
UOC: EMK, YNN 1
60 PAFZZ PAOZZ 5305-01-380-3395 80204 B18231B10025N . SCREW, CAP, HEX HEAD 1
61 PAFZZ PAOZZ 5310-01-099-9532 30554 88-22331-1 . WASHER, LOCK .... 1 62 XDFZZ XB 30554 88-22493 . FAN GUARD SUPPORT B LEFT
SIDE
UOC: EMK, YNN 1
63 PAFZZ PAOZZ 2930-01-370-2868 30554 88-20468 . TANK, RADIATOR, OVERFL (NOT
SHOWN) 1
64 XAFZZ XA 30554 88-20468-1 . . CAP 1
65 XAFZZ XA 30554 88-20468-2 . . BOTTLE 1
66 XAFZZ XA 30554 88-20468-3 . . CAGE, BOTTLE 1
67 XDFZZ XB 30554 88-22543 . BRACKET, OVERFLOW 1

END OF FIGURE

FIELD AND SUSTAINMENT MAINTENANCE

15 kW 50/60 AND 400 Hz SKID MOUNTED TACTICAL QUIET GENERATOR SETS

GROUP 05 COOLANT SYSTEM

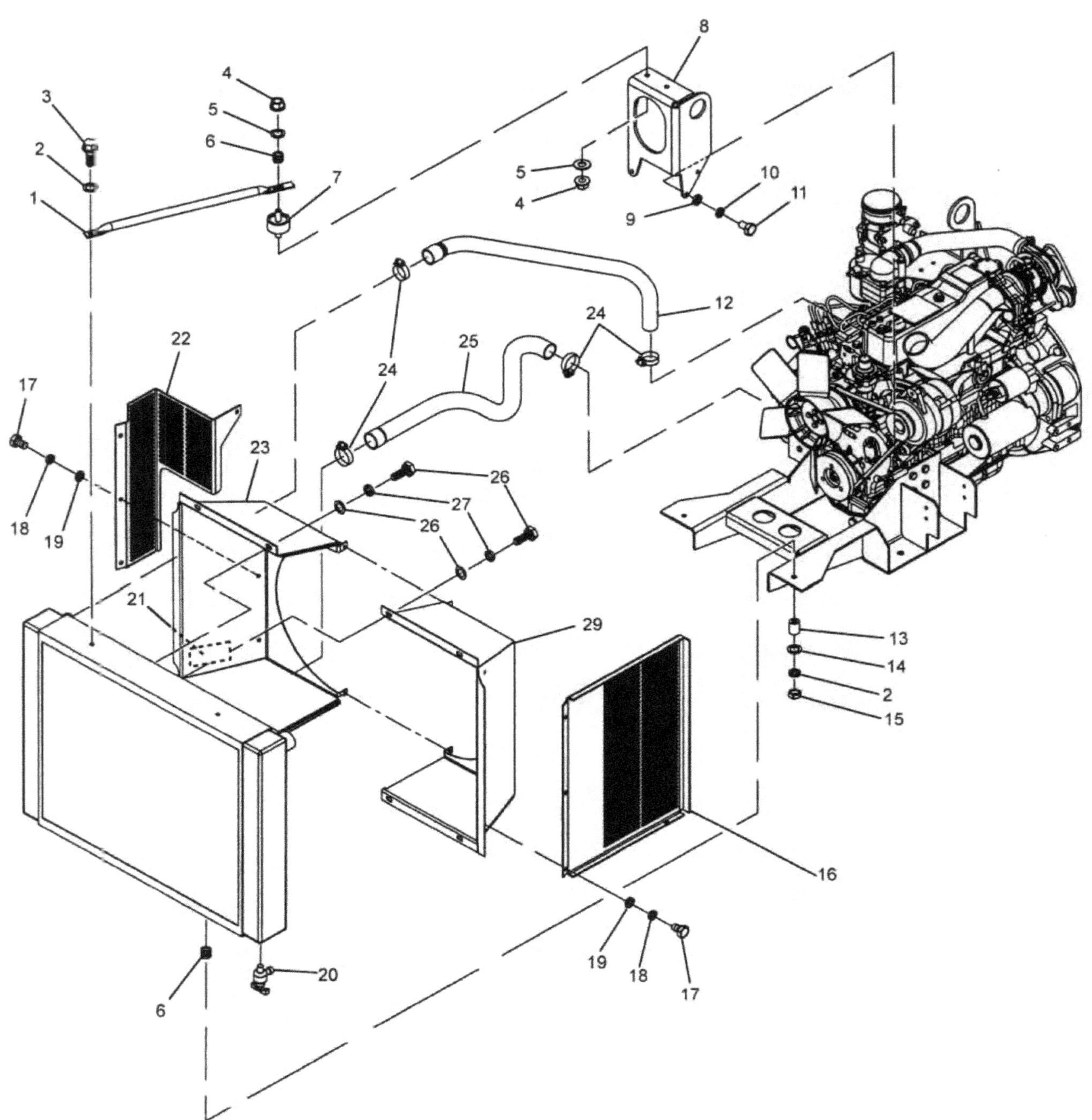

Figure 14. Coolant System.

(1) ITEM NO	(2) SMR CODE a. ARMY	b. AIR FORCE	(3) NSN	(4) CAGEC	(5) PART NUMBER	(6) DESCRIPTION AND USABLE ON CODE (UOC)	(7) QTY
						GROUP 05 COOLANT SYSTEM	
						FIG. 14 COOLANT SYSTEM	
1	XBFZZ	XBFZZ		30554	97-24034	. TIE-ROD, RADIATOR, UP UOC: 86Q, 86R	2
2	PAFZZ	PAFZZ	5310-01-531-6724	30554	88-20556-7	. WASHER, LOCK UOC: 86Q, 86R	4
3	PAFZZ	PAFZZ		30554	88-20260-51	. SCREW UOC: 86Q, 86R	2
4	PAFZZ	PAFZZ	5310-00-696-5173	30554	69-561-5	. NUT, PLAIN, ASSEMBLED UOC: 86Q, 86R	4
5	PAFZZ	PAFZZ	5310-00-809-4058	30554	88-20564-2	. WASHER, FLAT UOC: 86Q, 86R	2
6	PAFZZ	PAFZZ		30554	00-24003-1	. GROMMET, RADIATOR UOC: 86Q, 86R	2
7	XBFZZ	XBFZZ	5340-01-550-0921	30554	00-24004	. MOUNT, RADIATOR UOC: 86Q, 86R	2
8	XBFZZ	XBFZZ		30554	97-24019	. SUPPORT, RADIATOR, UPPER UOC: 86Q, 86R	1
9	PAFZZ	PAFZZ	5310-01-531-4937	30554	88-20033-24A	. WASHER, FLAT UOC: 86Q, 86R	3
10	PAFZZ	PAFZZ	5310-01-099-9532	30554	88-22331-1	. WASHER, LOCK UOC: 86Q, 86R	3
11	PAFZZ	PAFZZ	5306-01-179-1438	80204	B18231A10025N	. BOLT, MACHINE UOC: 86Q, 86R	3
12	PAFZZ	PAFZZ		30554	97-24036	. HOSE, UPPER, RADIATOR UOC: 86Q, 86R	1
13	PAFZZ	PAFZZ		30554	00-24003-2	. GROMMET, RADIATOR UOC: 86Q, 86R	2
14	PAFZZ	PAFZZ	5310-01-532-0321	30554	88-20033-31A	. WASHER, FLAT UOC: 86Q, 86R	2
15	PAFZZ	PAFZZ	5310-01-466-6687	30554	88-22790-3	. NUT, HEX UOC: 86Q, 86R	2
16	XBFZZ	XBFZZ		30554	97-24040	. GUARD, FAN, LEFT UOC: 86Q, 86R	1
17	PAFZZ	PAFZZ	5305-01-369-2166	30554	88-20260-30	. SCREW, CAP, HEX HEAD UOC: 86Q, 86R	6
18	PAFZZ	PAFZZ		30554	88-20556-38	. WASHER, LOCK UOC: 86Q, 86R	6
19	PAFZZ	PAFZZ	5310-01-531-4939	30554	88-20033-22A	. WASHER, FLAT UOC: 86Q, 86R	6
20	PAFZZ	PAFZZ	4820-01-381-5079	30554	88-22751	. COCK, DRAIN UOC: 86Q, 86R	1
21	PAFZZ	PAFZZ		30554	97-24146	. NAMEPLATE, OIL, CHECK UOC: 86Q, 86R	1
22	XBFZZ	XBFZZ		30554	97-24041	. GUARD, FAN, RIGHT UOC: 86Q, 86R	1
23	XBFZZ	XBFZZ		30554	97-24032	. SHROUD, RADIATOR RH UOC: 86Q, 86R	1
24	PAFZZ	PAFZZ	4730-01-470-1567	30554	88-20561-3	. CLAMP, HOSE UOC: 86Q, 86R	4
25	PAFZZ	PAFZZ		30554	97-24035	. HOSE, LOWER, RADIATOR UOC: 86Q, 86R	1

(1) ITEM NO	(2) SMR CODE a. ARMY	b. AIR FORCE	(3) NSN	(4) CAGEC	(5) PART NUMBER	(6) DESCRIPTION AND USABLE ON CODE (UOC)	(7) QTY
26	PAFZZ	PAFZZ	5305-01-369-2166	30554	88-20260-30	. SCREW, CAP, HEX HEAD UOC: 86Q, 86R	8
27	PAFZZ	PAFZZ	5310-01-531-4939	30554	88-20033-22A	. WASHER, FLAT UOC: 86Q, 86R	2
28	PAFZZ	PAFZZ		30554	88-20556-38	. WASHER, LOCK UOC: 86Q, 86R	2
29	XBFZZ	XBFZZ		30554	97-24033	. SHROUD, RADIATOR LH UOC: 86Q, 86R	1

END OF FIGURE

FIELD AND SUSTAINMENT MAINTENANCE

15 kW 50/60 AND 400 Hz SKID MOUNTED TACTICAL QUIET GENERATOR SETS

GROUP 06 FUEL SYSTEM: FUEL TANK, LINES, AND FITTINGS

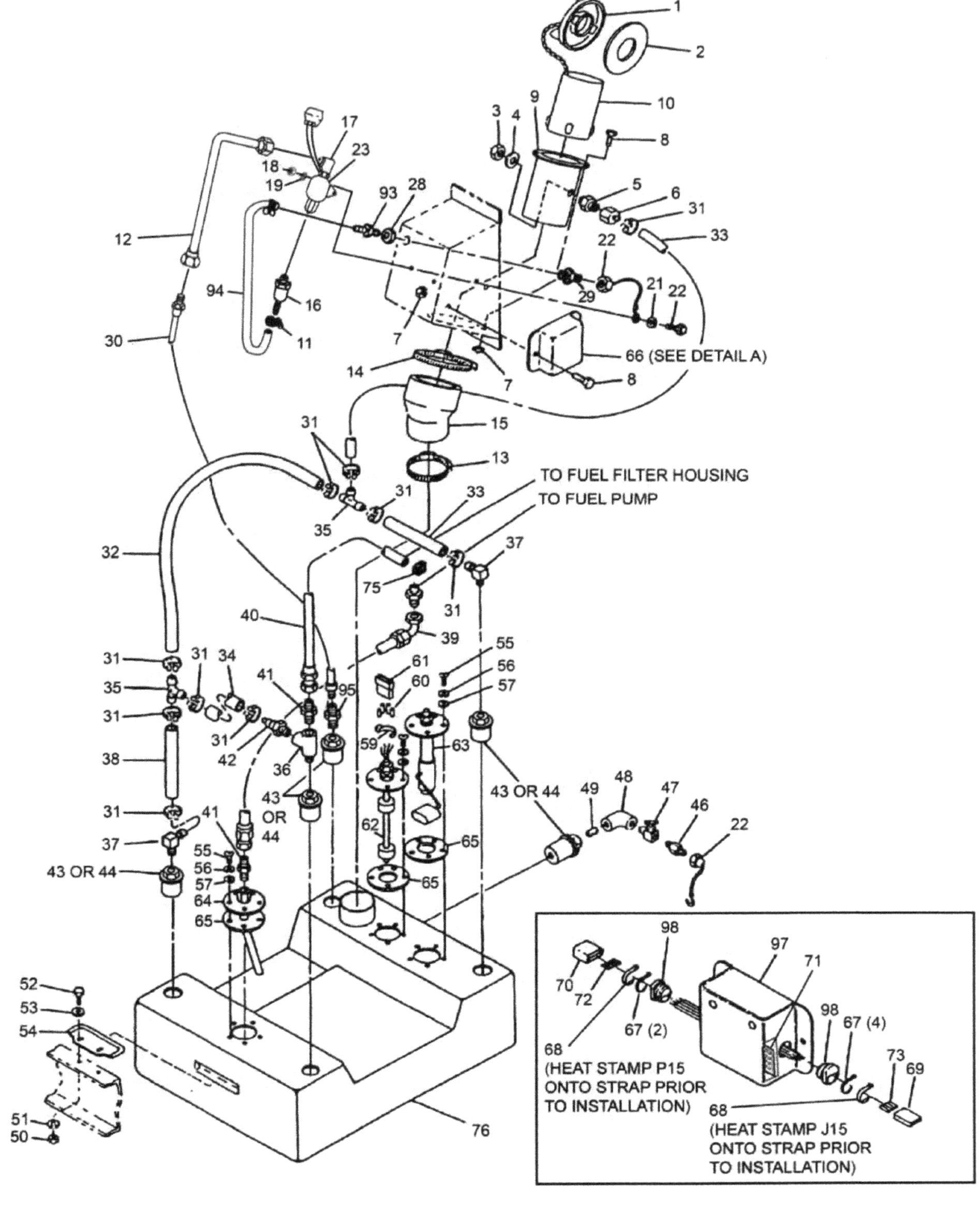

Figure 15. Fuel Tank, Lines, and Fittings (Sheet 1 of 2).

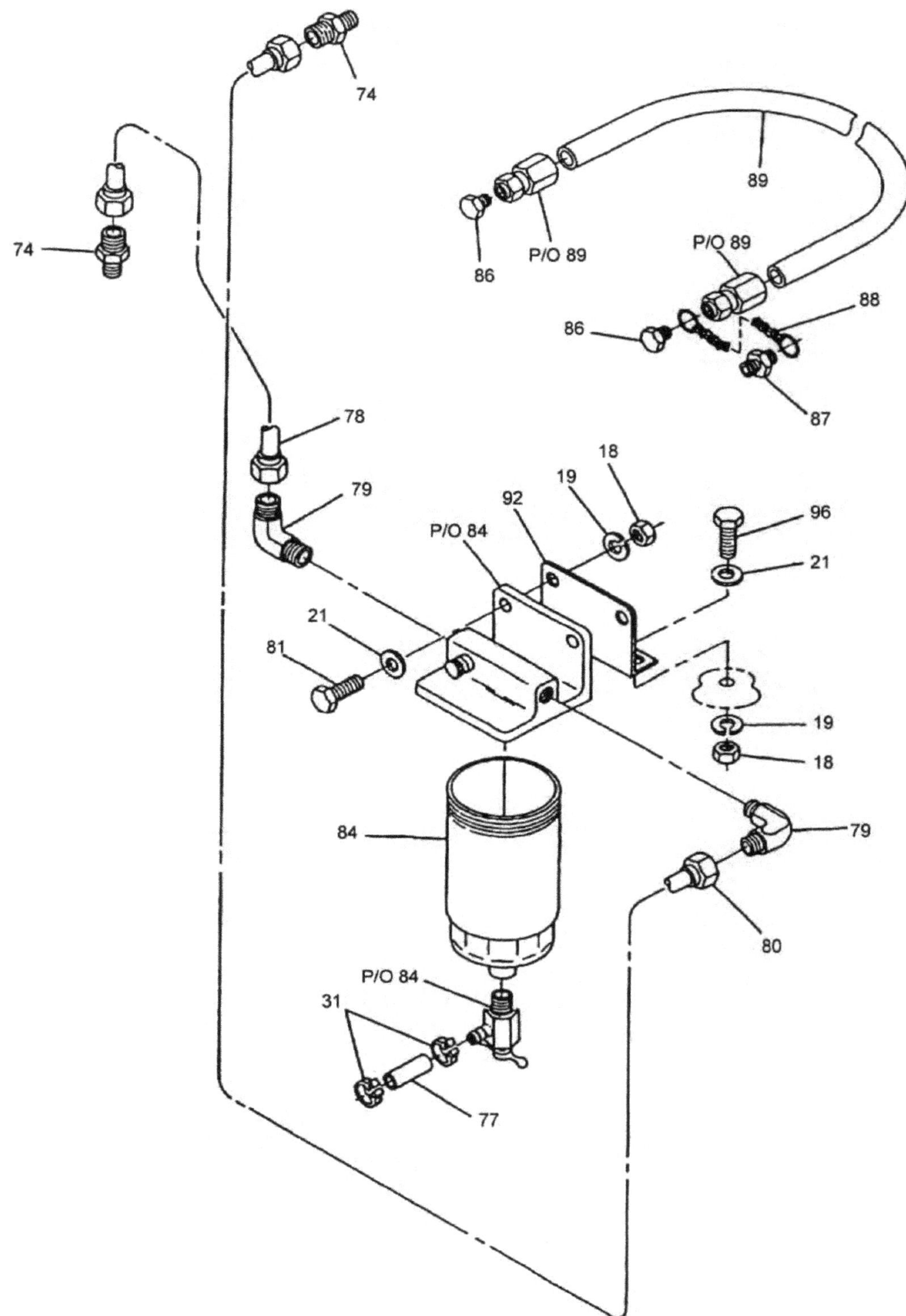

Figure 15. Fuel Tank, Lines, and Fittings (Sheet 2 of 2).

(1) (2) (3) (4) (5) (6) (7)
SMR CODE
a. b.
ITEM NO | ARMY AIR FORCE | NSN CAGEC PART NUMBER DESCRIPTION AND USABLE ON CODE (UOC) | QTY

GROUP 06 FUEL SYSTEM

FIG. 15 FUEL TANK, LINES, AND FITTINGS

1 PAFZZ PAOZZ 2590-00-141-9758 96906 MS35645-1 . CAP, FILLER OPENING 1
2 PAFZZ PAOZZ 5330-00-684-7851 96906 MS35643-1 . GASKET 1
3 PAFZZ PAOZZ 4730-01-532-0694 30554 88-22795 . NUT, SEAL 1
4 PAFZZ PAOZZ 5310-01-267-1685 96906 MS51412-8 . WASHER, FLAT 1
5 PAFZZ PAOZZ 4730-00-802-2560 81343 4-2 140137 . NIPPLE, HEX 1
6 PAFZZ PAOZZ 4730-01-134-9827 30554 88-21884 . CONNECTOR, FEMALE 1
7 PAFZZ PAOZZ 5310-01-012-3595 30554 69-561-6 . NUT, PLAIN, ASSEMBLED 8
8 PAFZZ PAOZZ 5306-01-156-7663 30554 88-20260-21 . SCREW, HEX WASHERHEAD 8
9 XDFZZ XB 5340-01-465-7761 30554 88-21892 . FILLER NECK, FUEL T 1
10 XDFZZ XB 5340-01-465-7765 30554 88-21893 . TUBE, FILLER NECK .... 1 11 PAFZZ PAOZZ 4730-00-908-3195 81343 SAE J1508-06 . CLAMP, HOSE
UOC: EMK, YNN 2
12 XDFZZ XB 30554 88-22111 . TUBE ASSY, OUTLET 1
13 PAFZZ PAOZZ 4730-01-470-1578 30554 88-20561-6 . CLAMP, HOSE 1
14 PAFZZ PAOZZ 4730-01-470-1701 30554 88-20561-7 . CLAMP, HOSE 1
15 PBFZZ PBOZZ 4720-01-392-0319 30554 88-22068 . HOSE, PREFORMED .... 1 16 PAFZZ PAOZZ 2910-01-553-6571 72850 00-24000 . FILTER, FUEL
UOC: EMK, YNN 1
17 PAFZZ PAOZZ 4730-00-187-0840 81343 4-2 070102C . ADAPTER, PIPE TO TUBE 1
18 PAFZZ PAOZZ 5310-01-466-6312 30554 88-22790-1 . NUT, PLAIN, HEX 6
19 PAFZZ PAOZZ 5310-00-274-8715 80205 MS35338-63 . WASHER, LOCK 6
20 PAFZZ PAOZZ 5306-00-484-5730 30554 88-20260-31 . BOLT, MACHINE 2
21 PAFZZ PAOZZ 5310-01-103-6042 96906 MS51412-4 . WASHER, FLAT 6
22 PAFZZ PAOZZ 4730-00-812-1333 93742 69-539-2 . CAP, TUBE .... 2 23 AFFFF AOO 30554 00-24002-1 . PUMP, FUEL
UOC: EMK, YNN 1
24 PAFZZ PAOZZ 5975-00-074-2072 96906 MS3367-1-9 . STRAP, TIEDOWN, ELECT
UOC: EMK, YNN 1
25 NOT USED
26 NOT USED
27 NOT USED
28 PAFZZ PAOZZ 4730-01-343-3192 96906 MS51860-54Z . LOCKNUT, TUBE FITTING
UOC: EMK, YNN 1
29 PAFZZ PAOZZ 4730-01-123-8618 96906 MS51520A5 . ADAPTER, STRAIGHT, TU
UOC: EMK, YNN 1
30 PAFZZ PAOZZ 4720-01-375-1391 30554 88-22338-1 . HOSE ASSEMBLY, NONMET 1
31 PAFZZ PAOZZ 4730-00-063-7919 30554 88-20029-4 . CLAMP, HOSE .... 12 32 MFFZZ MOO 30554 88-21604-146 . HOSE, NONMETALLIC MAKE FROM
P/N 88-20579-3 (30554), 33 INCHES REQUIRED
UOC: EMK 1
32 MFFZZ MOO 30554 97-24604-146 . HOSE, NONMETALLIC MAKE FROM
P/N 88-20579-3 (30554), 33 INCHES REQUIRED
UOC: 86Q.................................. 1

(1) ITEM NO	(2) SMR CODE a. ARMY	b. AIR FORCE	(3) NSN	(4) CAGEC	(5) PART NUMBER	(6) DESCRIPTION AND USABLE ON CODE (UOC)	(7) QTY
32	MFFZZ	MOO		30554	88-21605-146	. HOSE, NONMETALLIC MAKE FROM P/N 88-20579-3 (30554), 33 INCHES REQUIRED UOC: YNN	1
32	MFFZZ	MOO		30554	97-24605-146	. HOSE, NONMETALLIC MAKE FROM P/N 88-20579-3 (30554), 33 INCHES REQUIRED UOC: 86R	1
33	MFFZZ	MOO		30554	88-21604-104	. HOSE, NONMETALLIC MAKE FROM P/N 88-20579-3 (30554), 12 INCHES REQUIRED UOC: EMK	2
33	MFFZZ	MOO		30554	97-24604-104	. HOSE, NONMETALLIC MAKE FROM P/N 88-20579-3 (30554), 12 INCHES REQUIRED UOC: 86Q	2
33	MFFZZ	MOO		30554	88-21605-104	. HOSE, NONMETALLIC MAKE FROM P/N 88-20579-3 (30554), 12 INCHES REQUIRED UOC: YNN	2
33	MFFZZ	MOO		30554	97-24605-104	. HOSE, NONMETALLIC MAKE FROM P/N 88-20579-3 (30554), 12 INCHES REQUIRED UOC: 86R	2
34	MFFZZ	MOO		30554	88-21604-151	. HOSE, NONMETALLIC MAKE FROM P/N 88-20579-3 (30554), 20 INCHES REQUIRED UOC: EMK	1
34	MFFZZ	MOO		30554	97-24604-151	. HOSE, NONMETALLIC MAKE FROM P/N 88-20579-3 (30554), 20 INCHES REQUIRED UOC: 86Q	1
34	MFFZZ	MOO		30554	88-21605-151	. HOSE, NONMETALLIC MAKE FROM P/N 88-20579-3 (30554), 20 INCHES REQUIRED UOC: YNN	1
34	MFFZZ	MOO		30554	97-24605-151	. HOSE, NONMETALLIC MAKE FROM P/N 88-20579-3 (30554), 20 INCHES REQUIRED UOC: 86R	1
35	XDFZZ	XB	4730-01-474-2274	30554	88-21883	. TEE, HOSE	2
36	PAFZZ	PAOZZ	4730-00-177-6166	81343	4-4-4 140424C	. TEE, PIPE	1
37	PAFZZ	PAOZZ	4730-00-041-2526	30554	88-21114-2	. ELBOW, PIPE TO HOSE	1
38	MFFZZ	MOO		30554	88-21604-343	. HOSE, NONMETALLIC MAKE FROM P/N 88-20579-3 (30554), 9 INCHES REQUIRED UOC: EMK	1
38	MFFZZ	MOO		30554	97-24604-343	. HOSE, NONMETALLIC MAKE FROM P/N 88-20579-3 (30554), 9 INCHES REQUIRED UOC: 86Q	1
38	MFFZZ	MOO		30554	88-21605-343	. HOSE, NONMETALLIC MAKE FROM P/N 88-20579-3 (30554), 9 INCHES REQUIRED UOC: YNN	1
38	MFFZZ	MOO		30554	97-24605-343	. HOSE, NONMETALLIC MAKE FROM P/N 88-20579-3 (30554), 9 INCHES REQUIRED UOC: 86R	1

(1)	(2) SMR CODE a. ARMY	b. AIR FORCE	(3) NSN	(4) CAGEC	(5) PART NUMBER	(6) DESCRIPTION AND USABLE ON CODE (UOC)	(7) QTY

39 PAFZZ PAOZZ 4720-01-394-1931 30554 88-22326-1 . HOSE ASSEMBLY, NONME 1
40 PAFZZ PAOZZ 4720-01-366-7172 30554 88-22325 . HOSE ASSEMBLY, NONME 1
41 PAFZZ PAOZZ 4730-00-995-1559 96906 MS51500A6 . ADAPTER, STRAIGHT, PI 2
42 PAFZZ PAOZZ 4730-00-200-0531 30554 88-21173 . ADAPTER, STRAIGHT, PI 2
43 PAFZZ PAOZZ 4820-01-480-0846 30554 88-20043 . ADAPTER, DRAIN 5
44 PAFZZ PAOZZ 4930-01-475-0388 30554 88-20049 . FITTING ASSY .... 5 45 NOT USED
46 PAFZZ PAOZZ 4730-00-620-6904 96906 MS51500A5-4 . ADAPTER, STRAIGHT, PI 1
47 PAFZZ PAOZZ 4820-01-540-3024 30554 88-22096-2 . COCK, SHUTOFF, SCREW 1
48 PAFZZ PAOZZ 4730-00-613-6468 81343 4-4-130238 . ELBOW, PIPE 90 DEGREE 1
49 PAFZZ PAOZZ 4730-00-765-9103 30554 72-2173-1 . NIPPLE, PIPE 1
50 PAFZZ PAOZZ 5310-01-470-1286 30554 88-22790-2 . NUT, PLAIN, HEX 4
51 PAFZZ PAOZZ 5310-00-011-6120 96906 MS35338-64 . WASHER, LOCK 4
52 PAFZZ PAOZZ 5306-00-226-4827 80204 B1821BH03C100N . BOLT, MACHINE 4
53 PAFZZ PAOZZ 5310-00-044-6477 96906 MS51412-25 . WASHER, FLAT 4
54 XDFZZ XB 30554 88-21854 . HOLDDOWN, FUEL TANK 2
55 PAFZZ PAOZZ 5305-01-165-1254 96906 MS51492-14 . SCREW, MACHINE 15
56 PAFZZ PAOZZ 5310-00-274-8710 80205 MS35338-62 . WASHER, LOCK 15
57 PAFZZ PAOZZ 5310-01-234-9416 96906 MS51412-2 . WASHER, FLAT .... 15 58 AFFFF AOO 30554 88-22547 . SWITCH ASSY, FUEL SENDER (NOT
SHOWN) 1
59 PAFZZ PAOZZ 5975-00-074-2072 96906 MS3367-1-9 . . STRAP, TIEDOWN, ELECT 1
60 PAFZZ PAOZZ 5999-01-092-2655 30554 88-20476 . . CONTACT, ELECTRICAL 4
61 PAFZZ PAOZZ 5935-00-483-0259 30554 88-20475 . . CONNECTOR BODY, RECE 1
62 PAFZZ PAOZZ 2910-01-376-2268 30554 88-21121-1 . . SWITCH, FUEL LEVEL 1
63 PAFZZ PAOZZ 2910-01-369-5012 30554 88-21162-3 . SWITCH, FUEL LEVEL 1
64 XDFZZ XB 30554 88-21850 . PICK-UP TUBE ASSY 1
65 PAFZZ PAOZZ 5330-01-366-2836 30554 88-20286 . GASKET .... 3 66 PAFZZ PAOZZ 6680-01-392-8821 30554 88-20492 . FLOAT SWITCH, MODULE (NOT
SHOWN) 1
67 PAFZZ PAOZZ 5975-00-727-5153 96906 MS3367-4-9 . . STRAP, TIEDOWN, ELECT 6
68 PAFZZ PAOZZ 5975-00-074-2072 96906 MS3367-1-9 . . STRAP, TIEDOWN, ELECT 2
69 PAFZZ PAOZZ 5935-00-315-9563 30554 88-20474 . . CONNECTOR BODY, RECE 1
70 PAFZZ PAOZZ 5935-00-483-0259 30554 88-20475 . . CONNECTOR BODY, RECE .... 1 71 XAFFF XAOOO 30554 88-20493 . . PC BOARD ASSY, FLO (NOT
SHOWN) 1
72 PAFZZ PAOZZ 5999-01-092-2655 30554 88-20476 ...CONTACT,ELECTRICAL 4
73 PAFZZ PAOZZ 5999-01-039-8438 30554 88-20477 ...CONTACT,ELECTRICAL .... 4 74 PAFZZ PAOZZ 4730-01-376-4256 30554 88-22324 . ADAPTER, STRAIGHT, TUBE
UOC: EMK, YNN 3
75 PAFZZ PAOZZ 4730-00-024-3971 81343 SAE J1508 . CLAMP, HOSE 1
76 PBFZZ PBOZZ 2910-01-388-6383 30554 88-21043 . TANK, FUEL, ENGINE .... 1 77 MFFZZ MOO 30554 88-21604-339 . HOSE, NONMETALLIC MAKE FROM
P/N 88-20579-3 (30554), 32 INCHES REQUIRED
UOC: EMK 1
77 MFFZZ MOO 30554 97-24604-339 . HOSE, NONMETALLIC MAKE FROM
P/N 88-20579-3 (30554), 32 INCHES REQUIRED
UOC: 86Q.................................. 1
77 MFFZZ MOO 30554 88-21605-339 . HOSE, NONMETALLIC MAKE FROM
P/N 88-20579-3 (30554), 32 INCHES REQUIRED
UOC: YNN 1

(1)	(2) SMR CODE a.	b.	(3)	(4)	(5)	(6)	(7)
ITEM NO	ARMY	AIR FORCE	NSN	CAGEC	PART NUMBER	DESCRIPTION AND USABLE ON CODE (UOC)	QTY

77 MFFZZ MOO 30554 97-24605-339 . HOSE, NONMETALLIC MAKE FROM
P/N 88-20579-3 (30554), 32 INCHES
REQUIRED
UOC: 86R 1

78 XDFZZ XB 30554 88-22486 . TUBE, FUEL FILTER
UOC: EMK, YNN 1

79 PAFZZ PAOZZ 4730-00-812-7999 96906 MS51504A6 . ELBOW, PIPE TO TUBE
UOC: EMK, YNN 2

80 XDFZZ XB 30554 88-22485 . TUBE, FILTER INLET
UOC: EMK, YNN 1

81 PAFZZ PAOZZ 5305-00-225-3843 80204 B1821BH025C100N . SCREW, CAP, HEX HEAD 4

82 NOT USED

83 NOT USED

84 PBFFF PBOOO 2910-01-364-9843 30554 88-21101 . FILTER, FUEL
UOC: EMK, YNN 1

85 PAFZZ PAOZZ 4720-00-021-3320 30554 69-668 . LINE, FUEL, AUXILIARY (NOT
SHOWN) 1

86 PAFZZ PAOZZ 5340-01-086-2049 30554 88-20024-5 . . PLUG, TUBING 2

87 PAFZZ PAOZZ 4730-01-126-2173 30554 75-0809 . . UNION .... 1 88 MFFZZ MOO 30554 69-668-2 . . CHAIN, WELDLESS MAKE FROM P/
N RR-C-271 TY2CL6 (80244), 13
LINKS REQD 1

89 PAFZZ PAOZZ 4720-01-215-0816 30554 88-20551-1 . . HOSE ASSEMBLY, NONME 1

90 NOT USED

91 NOT USED

92 XDFZZ XB 30554 88-22748 . BRACKET, SUPPORT
UOC: EMK, YNN 1

93 PAFZZ PAOZZ 4730-01-407-0649 30554 88-22765 . ADAPTER, STRAIGHT, TUBE
UOC: EMK, YNN 1

94 MFFZZ MOO 30554 88-21604-396 . HOSE, NONMETALLIC MAKE FROM
P/N 88-20579-4 (30554), 11 INCHES
REQUIRED
UOC: EMK 1

94 MFFZZ MOO 30554 88-21605-396 . HOSE, NONMETALLIC MAKE FROM
P/N 88-20579-4 (30554), 11 INCHES
REQUIRED
UOC: YNN 1

95 PAFZZ PAOZZ 4730-01-309-6370 96906 MS51500B4-4Z . ADAPTER, STRAIGHT, PI 1

96 PAFZZ PAOZZ 5305-00-068-0508 80204 B1821BH025C075N . SCREW, CAP, HEX HEAD 2

97 PAFZZ PAOZZ 30554 88-22643 . . CHASSIS ASSY, FL SW MOD 1

98 PAFZZ PAOZZ 30554 88-22654 . . BUSHING, STRAIN RELIEF 2

END OF FIGURE

FIELD AND SUSTAINMENT MAINTENANCE

15 kW 50/60 AND 400 Hz SKID MOUNTED TACTICAL QUIET GENERATOR SETS

GROUP 06 FUEL SYSTEM: FUEL TANK, LINES, AND FITTINGS

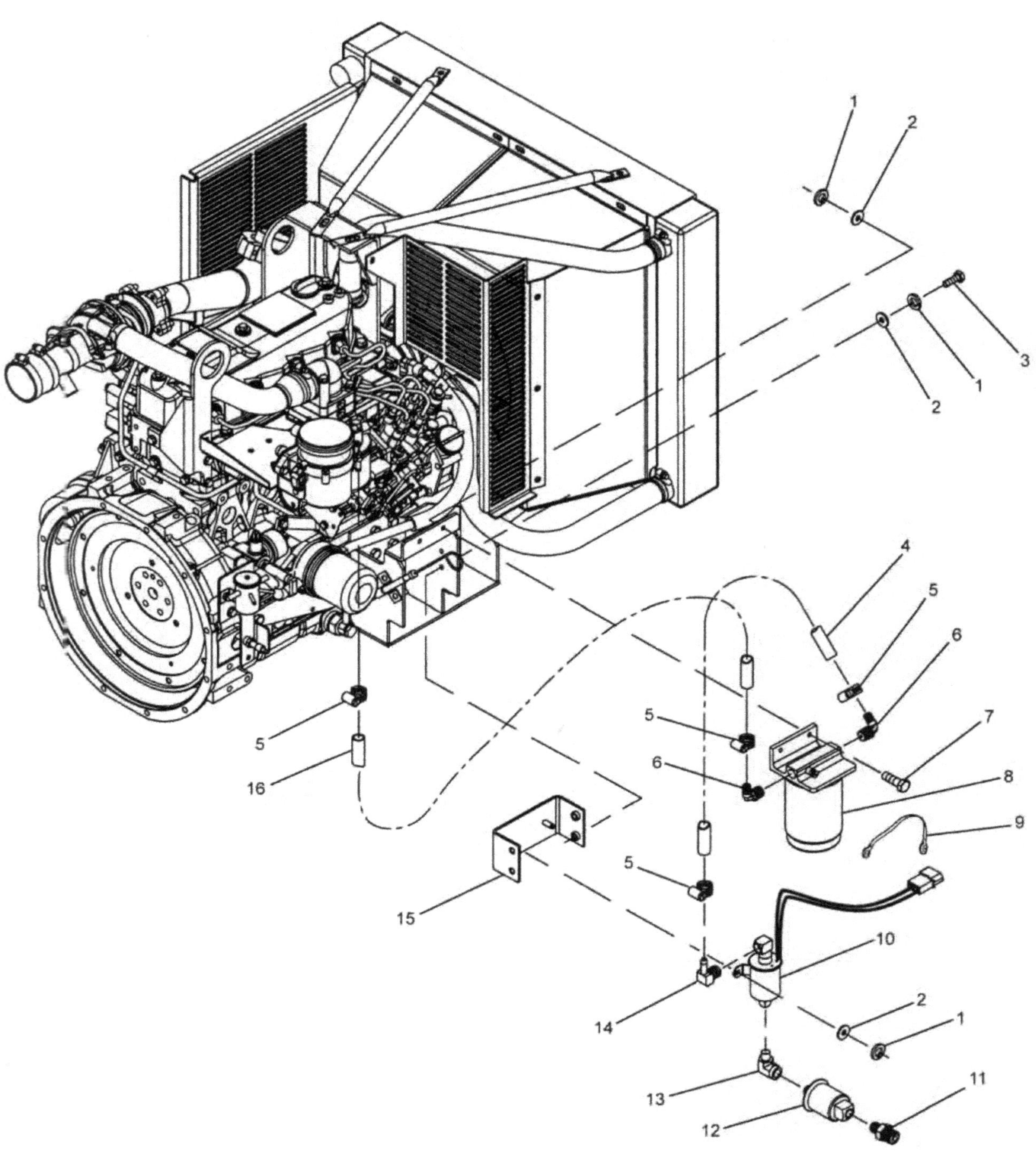

Figure 16. Fuel Tank, Lines, and Fittings.

(1) ITEM NO	(2) SMR CODE a. ARMY	b. AIR FORCE	(3) NSN	(4) CAGEC	(5) PART NUMBER	(6) DESCRIPTION AND USABLE ON CODE (UOC)	(7) QTY
						GROUP 06 FUEL SYSTEM	
						FIG. 16 FUEL TANK, LINES, AND FITTINGS	
1	PAFZZ	PAFZZ		30554	88-20556-38	. WASHER, LOCK UOC: 86Q, 86R	10
2	PAFZZ	PAFZZ	5310-01-531-4939	30554	88-20033-22A	. WASHER, FLAT UOC: 86Q, 86R	10
3	PAFZZ	PAFZZ	5305-01-369-2166	30554	88-20260-30	. SCREW, CAP, HEX HEAD UOC: 86Q, 86R	4
4	MFFZZ	MOF		30554	97-24071	. HOSE, 0.75 DIA., 13.0 IN. MAKE FROM (98411) 208-5 UOC: 86Q, 86R	1
5	PAFZZ	PAFZZ	4730-01-470-1626	30554	88-20561-1	. CLAMP, HOSE UOC: 86Q, 86R	4
6	PAFZZ	PAFZZ	4730-01-051-9840	81343	4-4 070202C	. ELBOW UOC: 86Q, 86R	2
7	PAFZZ	PAFZZ	5305-01-379-5734	30554	88-20260-35	. SCREW/WASHER UOC: 86Q, 86R	2
8	PAFZZ	PAFZZ	2910-01-364-9843	30554	88-21101	. FILTER, FLUID UOC: 86Q, 86R	2
9	PAFZZ	PAFZZ	5975-00-074-2072	30554	88-20018-1	. STRAP, TIEDOWN, ELECTR UOC: 86Q, 86R	1
10	PAFZZ	PAFZZ		30554	88-24002-1	. PUMP, FUEL UOC: 86Q, 86R	1
11	PAFZZ	PAFZZ	4730-00-187-0840	30554	88-21755-4	. ADAPTER, PIPE TO TUBE UOC: 86Q, 86R	1
12	PAFZZ	PAFZZ	2910-01-553-6571	30554	00-24000	. FILTER, FUEL UOC: 86Q, 86R	1
13	PAFZZ	PAFZZ	4730-00-725-3664	81343	2-2 140239C	. ELBOW UOC: 86Q, 86R	1
14	PAFZZ	PAFZZ		30554	98-19744-03	. FITTING, BEADED TUBE UOC: 86Q, 86R	1
15	XBFZZ	XBFZZ		30554	97-24044	. BRACKET, FUEL PUMP UOC: 86Q, 86R	3
16	MFFZZ	MOF		30554	97-24071	. HOSE, 0.75 DIA., 13.0 IN. MAKE FROM (98441) 208-5 UOC: 86Q, 86R	1

END OF FIGURE

FIELD AND SUSTAINMENT MAINTENANCE

15 kW 50/60 AND 400 Hz SKID MOUNTED TACTICAL QUIET GENERATOR SETS
GROUP 07 OUTPUT BOX ASSEMBLY: OUTPUT BOX INSTALLATION

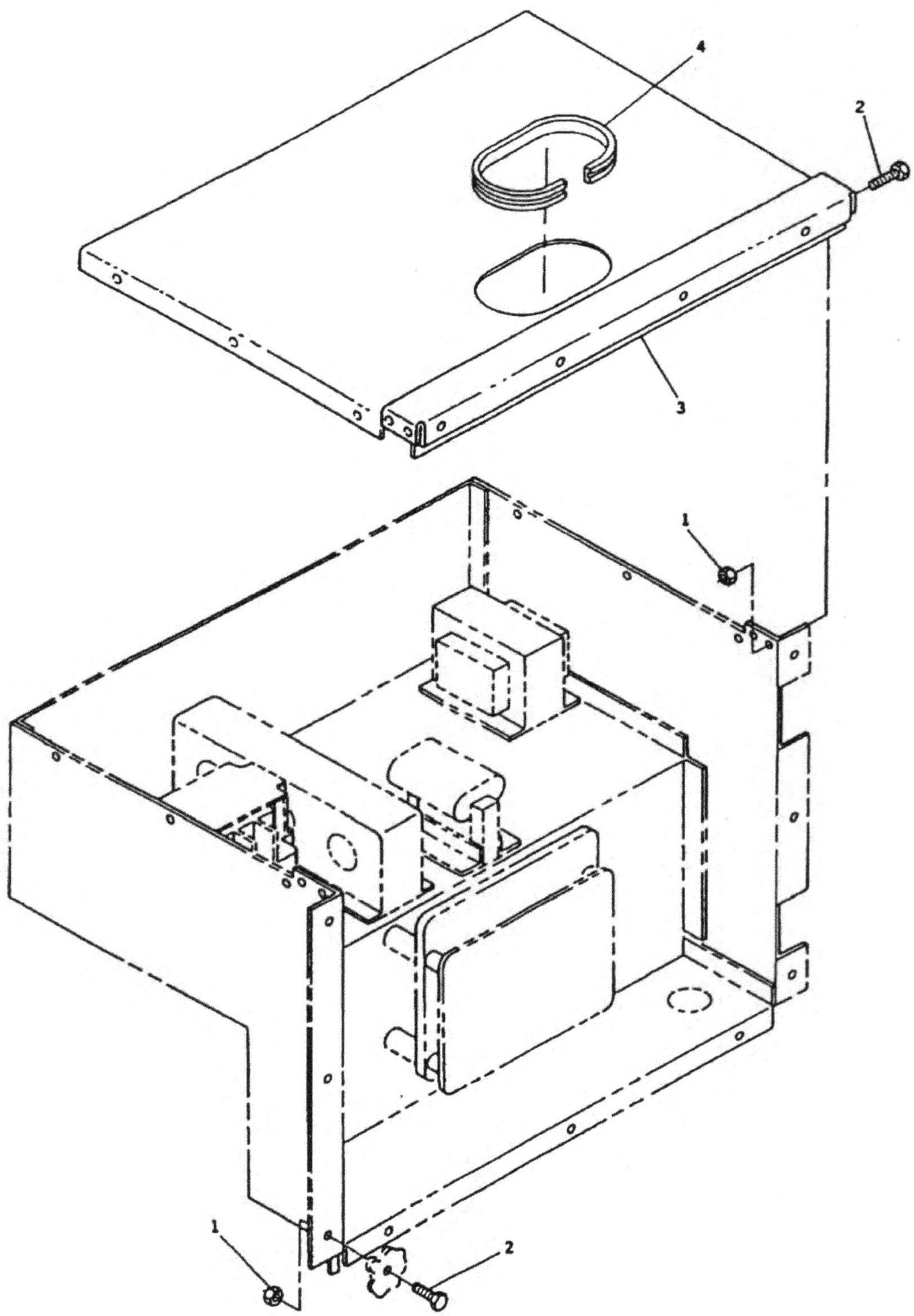

Figure 17. Output Box Installation.

(1)	(2) SMR CODE a. ARMY	b. AIR FORCE	(3) NSN	(4) CAGEC	(5) PART NUMBER	(6) DESCRIPTION AND USABLE ON CODE (UOC)	(7) QTY
ITEM NO							
						GROUP 07 OUTPUT BOX ASSEMBLY	
						FIG. 17 OUTPUT BOX INSTALLATION	
1	PAFZZ	PAOZZ	5310-01-012-3595	30554	69-561-6	. NUT, PLAIN, ASSEMBLED	19
2	PAFZZ	PAOZZ	5306-01-156-7663	30554	88-20260-21	. SCREW, HEX WASHERHEAD	19
3	XDFZZ	XB		30554	88-22701	. OUTPUT BOX, TOP	1
4	MFFZZ	MOO		30554	88-22704	. EDGING, RUBBER MAKE FROM P/N ZX-4134 (1X968), AS REQUIRED	1

END OF FIGURE

FIELD AND SUSTAINMENT MAINTENANCE

15 kW 50/60 AND 400 Hz SKID MOUNTED TACTICAL QUIET GENERATOR SETS

GROUP 07 OUTPUT BOX ASSEMBLY: ENGINE/OUTPUT BOX HARNESS

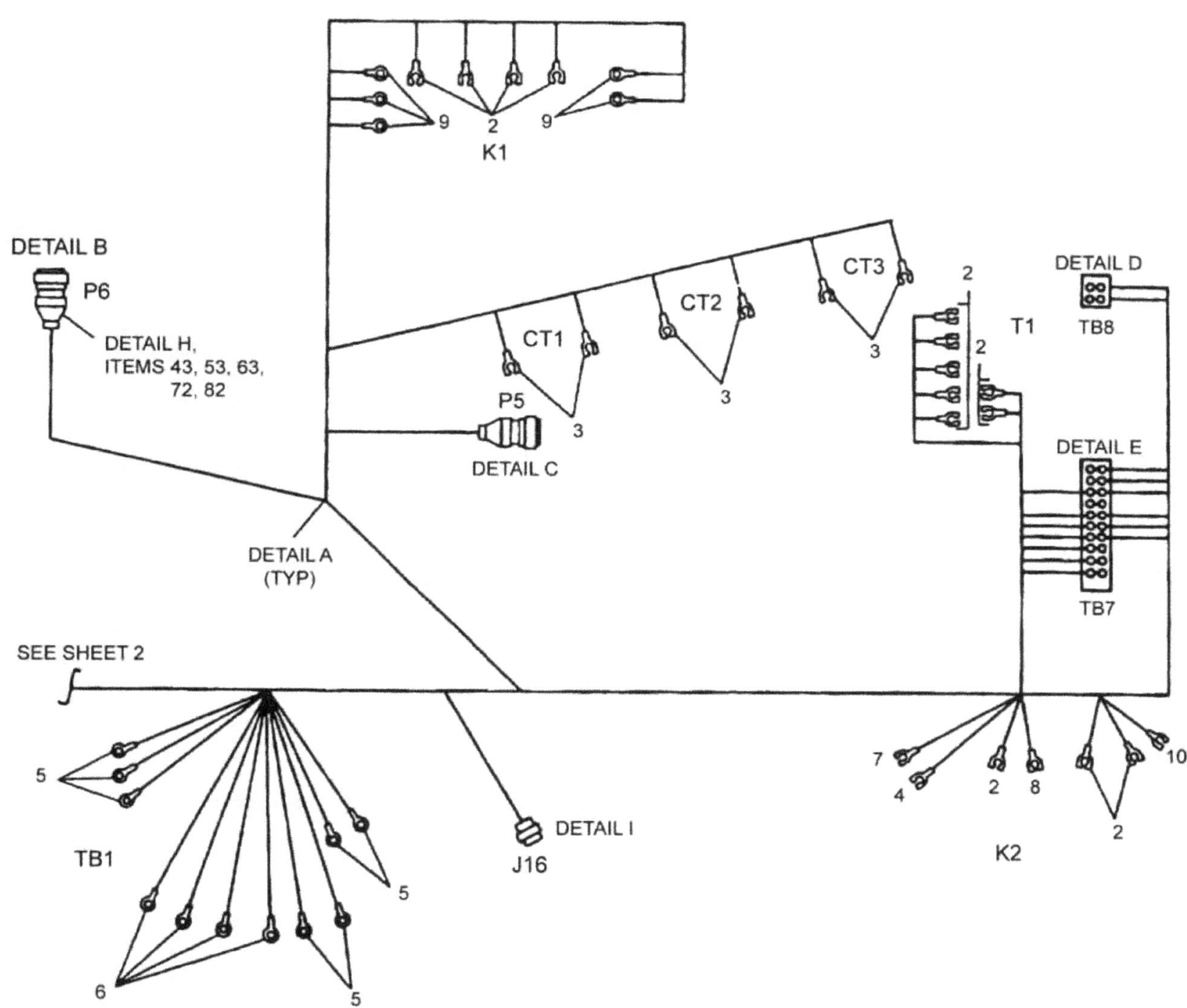

Figure 18. Engine/Output Box Harness (Sheet 1 of 8).

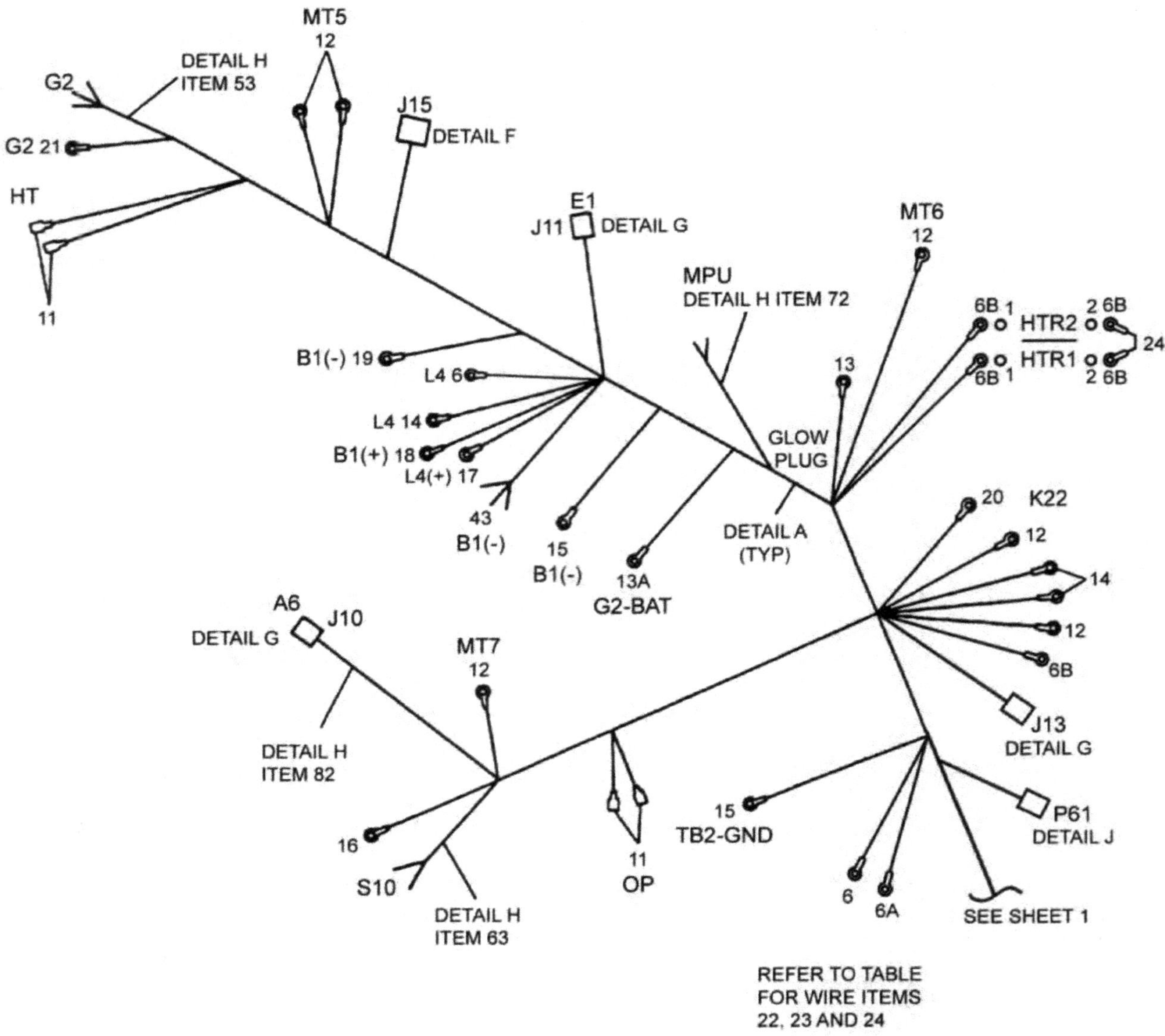

Figure 18. Engine/Output Box Harness (Sheet 2 of 8).

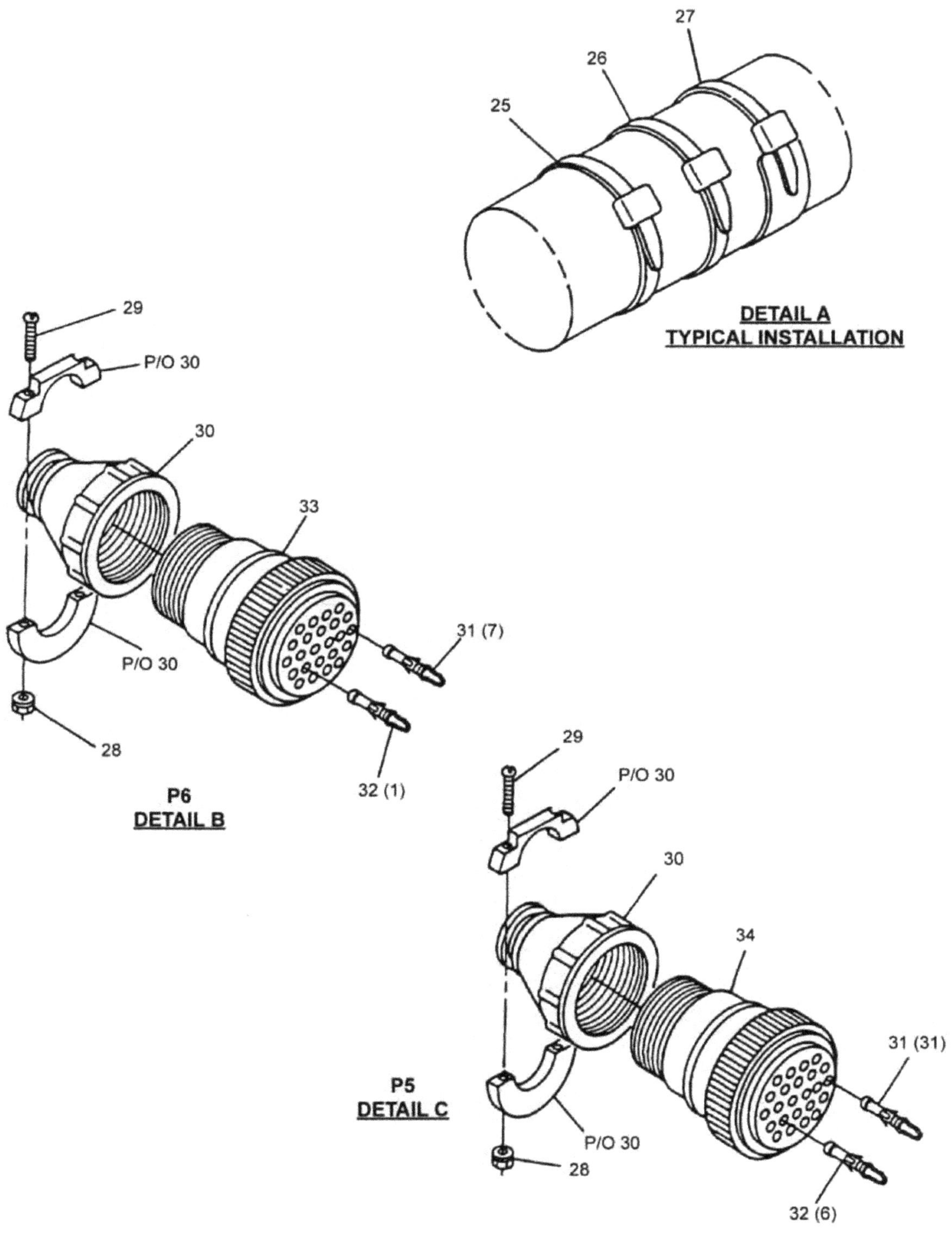

Figure 18. Engine/Output Box Harness (Sheet 3 of 8).

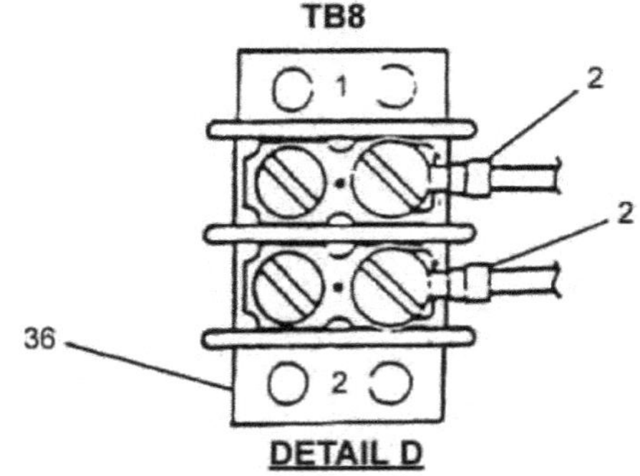

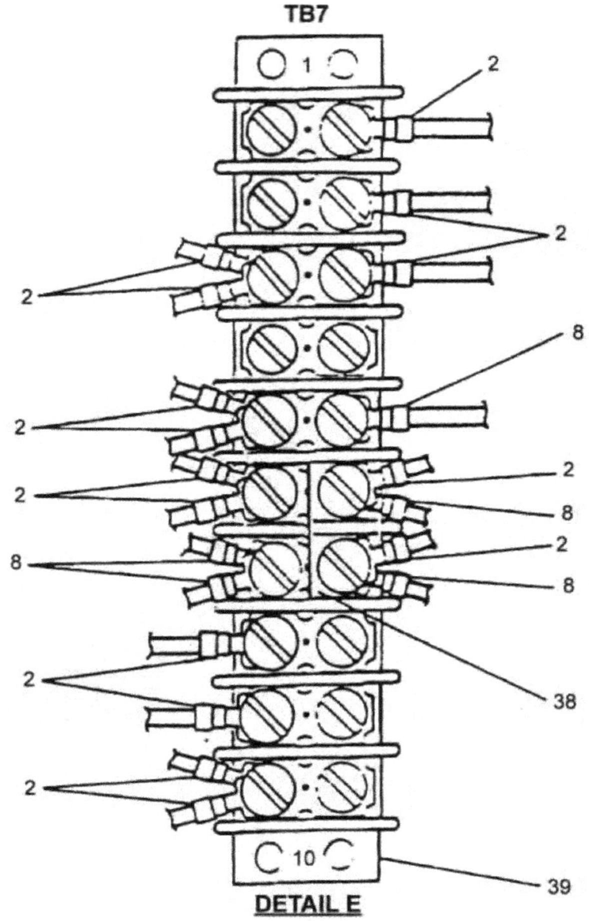

Figure 18. Engine/Output Box Harness (Sheet 4 of 8).

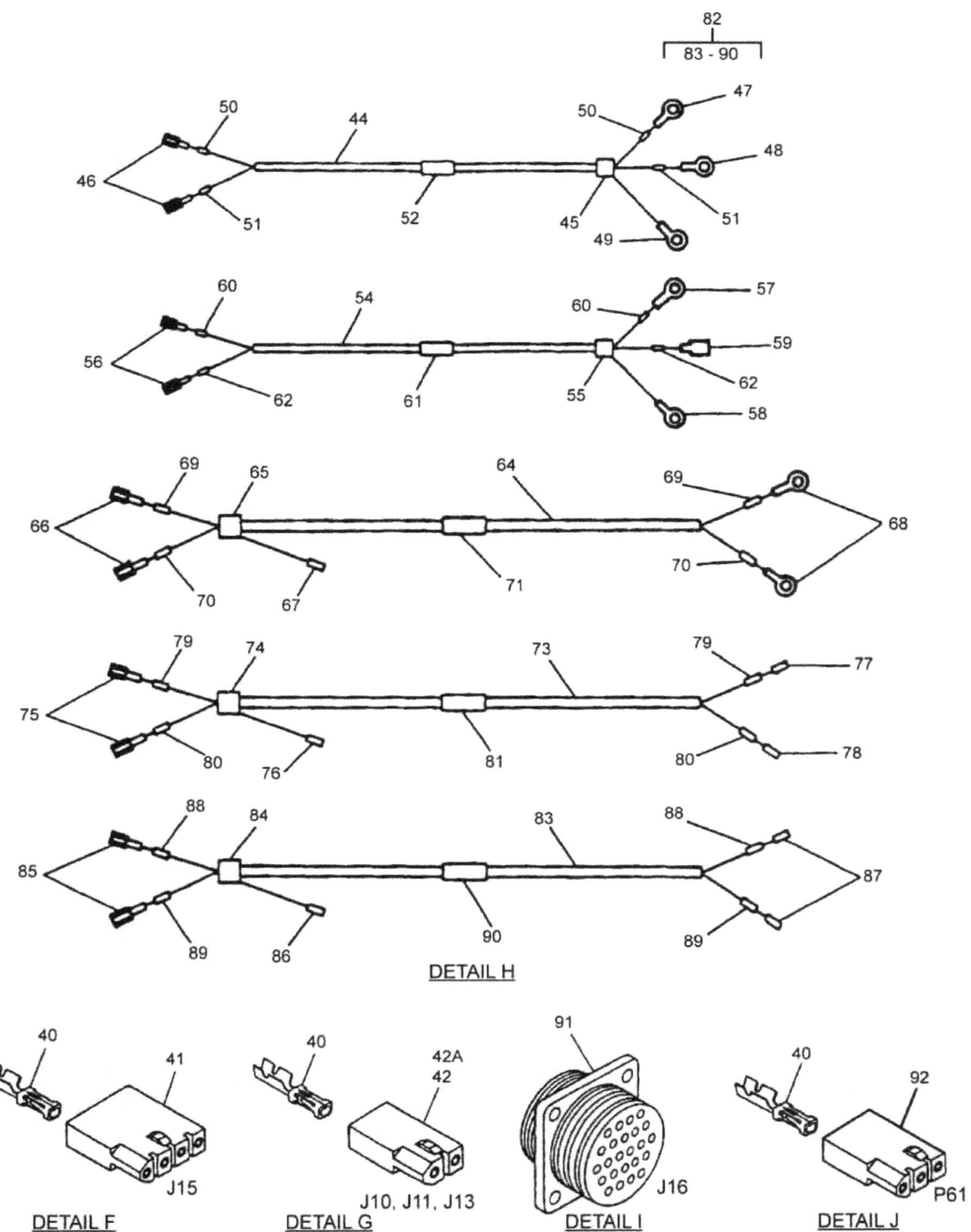

Figure 18. Engine/Output Box Harness (Sheet 5 of 8).

WIRE REF NO.	WIRE MARKING NO.	FROM	FIG.8 ITEM NO.	TO	FIG.8 ITEM NO.	FIG.8 ITEM NO.	MARKING COLOR	WIRE LENGTH
1	100AU20	MT5-GND	12	TB7-6	2	35	RED	76.0
2	100AW20	K2-X2	2	TB7-10	2	35	RED	7.0
3	100AX16	J16-E	–	TB7-7	8	37	BLACK	36.0
4	100BA20	K22-2	12	TB7-10	2	35	RED	31.5
5	NOT USED							
6	NOT USED							
7	NOT USED							
8	100C10	TB2-GND	15	B1 (–)	15	24	BLACK	41.0
9	100K20	K1-11	2	TB7-6	2	35	RED	44.0
10	100L16	B1 (–)	19	TB7-7	8	37	RED	34.0
11	100M20	J11-2	40	TB7-6	2	35	RED	58.0
12	100P20	J15-4	40	TB7-7	2	35	RED	72.0
13	100R16	G2-E	21	TB7-6	8	37	RED	64.0
14	100S16	K2-X2	8	TB7-7	8	37	RED	12.0
15	101B20	P5-1	31	K1-A1	9	35	BLACK	14.0
16	102B20	P5-2	31	K1-B1	9	35	BLACK	15.0
17	103B20	P5-3	31	K1-C1	9	35	BLACK	16.0
18	185L20	P5-4	31	K1-X	2	35	RED	19.0
19	105A20	P5-5	31	K1-12	2	35	RED	24.5
20	106A20	P5-6	31	TB1-6	5	35	BLACK	28.0
21	107B16	P5-36	32	TB1-7	6	37	BLACK	33.0
22	107C20	TB1-7	5	T1-2	2	35	BLACK	31.0
23	107L16	P5-37	32	TB1-7	6	37	BLACK	33.0
24	108A20	P5-8	31	TB1-8	5	35	BLACK	30.0
25	108D20	TB1-8	5	TB7-8	2	35	BLACK	24.0
26	109A20	P5-9	31	TB1-9	5	35	BLACK	28.0
27	109J20	T1-1	2	TB1-9	5	35	BLACK	27.5
28	110B20	TB1-11	5	TB7-9	2	35	BLACK	24.0
29	110C16	P5-10	32	TB1-12	6	37	BLACK	30.0
30	110D16	P5-11	32	TB1-12	6	37	BLACK	30.0
31	111A20	P5-12	31	CT1-A1	3	35	BLACK	12.0
32	112A20	P5-13	31	CT1-A2	3	35	BLACK	13.0
33	113A20	P5-14	31	CT2-B1	3	35	BLACK	15.0
34	114A20	P5-15	31	CT2-B2	3	35	BLACK	17.0
35	115A20	P5-16	31	CT3-C1	3	35	BLACK	18.0
36	116A20	P5-17	31	CT3-C2	3	35	BLACK	21.0
37	NOT USED							
38	122B20	P5-19	31	K1-B2	9	35	BLACK	25.0
39	123B20	P5-20	31	K1-C2	9	35	BLACK	25.0
40	125A20	P6-23	31	B1 (+)	18	35	RED	52.0
41	NOT USED							
42	131C20	P5-22	31	T1-3	2	35	BLACK	35.0
43	132C20	P5-23	31	T1-4	2	35	BLACK	35.0
44	133C20	P5-24	31	T1-5	2	35	BLACK	34.5
45	135E20	P5-25	31	TB7-1	2	35	BLACK	30.0
46	138C20	P5-26	31	T1-6	2	35	BLACK	34.0
47	139C20	P5-27	31	T1-7	2	35	BLACK	33.5

Figure 18. Engine/Output Box Harness (Sheet 6 of 8).

WIRE REF NO.	WIRE MARKING NO.	FROM	FIG.8 ITEM NO.	TO	FIG.8 ITEM NO.	FIG.8 ITEM NO.	MARKING COLOR	WIRE LENGTH
48	140D20	P5-28	31	TB8-1	2	35	BLACK	31.0
49	141D20	P5-29	31	TB8-2	2	35	BLACK	30.5
50	142C20	P5-30	31	TB7-2	2	35	BLACK	29.0
51	NOT USED							
52	165E12	K22-3	20	K2-A1	10	23	RED	36.5
53	165F8	K22-3	14	L4-1	14	22	RED	40.0
54	NOT USED							
55	166A16	K2-A2	7	L4 (+)	17	37	RED	38.0
56	166B20	P6-22	31	K2-A2	4	35	RED	22.0
57	167A20	K2-X1	2	S10-1	16	35	RED	81.0
58	167B20	P6-21	31	K2-X1	2	35	RED	23.0
59	168A20	OP-C	11	TB7-5	2	35	RED	82.0
60	168B20	HT-C	11	TB7-5	2	35	RED	85.0
61	168C16	P6-5	32	TB7-5	8	37	RED	28.0
62	169A20	P6-20	31	OP-NC	11	35	RED	82.0
63	171D20	P6-19	31	J15-3	40	35	RED	74.0
64	172D20	P6-18	31	K22-1	12	35	RED	44.0
65	173B20	P5-31	31	J15-2	40	35	RED	75.0
66	174A20	J15-1	40	TB7-3	2	35	RED	70.0
67	174B20	J11-1	40	TB7-3	2	35	RED	59.0
68	174C20	P5-32	31	TB7-3	2	35	RED	29.0
69	175B20	P5-33	31	MT5	12	35	RED	84.0
70	176B20	P5-34	31	MT6	12	35	RED	55.0
71	177B20	P5-35	31	MT7	12	35	RED	77.0
72	NOT USED							
73	223B20	P6-6	31	HT-NO	11	35	RED	41.0
74	191B16	P5-18	32	J16-B	–	37	BLACK	24.0
75	193B16	P5-21	32	J16-A	–	37	BLACK	24.0
76	NOT USED							
77	240A20	P5-7	31	K1-Y	2	35	RED	22.5
78	243A8	K22-4	14	GLO- PLUG (EMK, YNN)	13	22	RED	17.5
			USABLE ON 86Q, 86R ONLY					
78	243A8	K22-4	14	HTR1-1	6B	22	RED	17.5
79	NOT USED							
80	225P16	P61-1	40	J13-1	40	37	RED	40.0
81	100FP16	J13-2	40	ENG GND	6	37	RED	25.0
82	164M16	P61-2	40	G2-BAT	13A	37	RED	26.0
83	165S16	P61-3	40	L4-1	6	37	RED	48.5
84	JUMPER	HTR1-2	6B	HTR2-2	6B	22	N/A	4.0
85	100HTR	HTR2-1	6B	ENG GND	6A	22	RED	21.5

Figure 18. Engine/Output Box Harness (Sheet 7 of 8).

SHIELDED CABLE CONNECTION TABLE						
CABLE ITEM NO.	WIRE	FROM	FIG.8 ITEM NO.	TO	FIG.8 ITEM NO.	WIRE LENGTH
43	(SHIELD)	(CUTOFF)	-	B1 (-)	49	59.0
	100D16	P6-1	46	B1 (-)	47	59.0
	165C16	P6-2	46	L4-1	48	59.0
53	(SHIELD)	(CUTOFF)	-	G2-E	58	76.0
	164D16	P6-4	56	G2-BAT	57	76.0
	248A16	P6-3	56	G2-R	59	76.0
63	(SHIELD)	P6-9	67	(CUTOFF)	-	55.0
	163D16	P6-7	66	S10-2	68	55.0
	178A16	P6-8	66	S10-3	68	55.0
72	(SHIELD)	P6-12	76	(CUTOFF)	-	49.0
	147B16	P6-10	75	MPU-1	78	49.0
	148B16	P6-11	75	MPU-2	77	49.0
82	(SHIELD)	P6-15	86	(CUTOFF)	-	50.0
	149B16	P6-13	85	J10-1	87	50.0
	150B16	P6-14	85	J10-2	87	50.0

Figure 18. Engine/Output Box Harness (Sheet 8 of 8).

(1)	(2) SMR CODE a.	b.	(3)	(4)	(5)	(6)	(7)
ITEM NO	ARMY	AIR FORCE	NSN	CAGEC	PART NUMBER	DESCRIPTION AND USABLE ON CODE (UOC)	QTY

GROUP 07 OUTPUT BOX ASSEMBLY

FIG. 18 ENGINE/OUTPUT BOX HARNESS

1 PBFFF PBOOO 5340-00-066-1235 30554 88-22164 . . WIRING HARNESS, BRAN (NOT SHOWN)
UOC: EMK, YNN 1
1 PBFFF PBOOO 5995-01-571-2043 30554 97-24048 . . WIRING HARNESS, BRAN (NOT SHOWN)
UOC: 86Q, 86R............................ 1
2 PAFZZ PAOZZ 5940-01-425-2020 30554 88-20274-1 ...TERMINAL, SPADE #6, 22-18 AWG 31
3 PAFZZ PAOZZ 5940-01-531-6448 30554 88-20274-3 ...TERMINAL, SPADE #8, 22-18 AWG 6
4 PAFZZ PAOZZ 5940-01-369-2270 30554 88-20274-4 ...TERMINAL, SPADE #10, 22-18 AWG 1
5 PAFZZ PAOZZ 5940-00-143-4773 96906 MS25036-105 ...TERMINAL, LUG 7
6 PAFZZ PAOZZ 5940-00-143-4793 96906 MS25036-110 ...TERMINAL, LUG .... 6 6 PAFZZ PAOZZ 5940-00-114-1314 96906 MS20659-129 ...TERMINAL, LUG
UOC: 86Q, 86R............................ 1
6 PAFZZ PAOZZ 5940-00-114-1305 96906 MS25036-116 ...TERMINAL, LUG
UOC: 86Q, 86R............................ 4
7 PAFZZ PAOZZ 5940-01-367-9569 30554 88-20274-6 ...TERMINAL, SPADE #10, 16-14 AWG 1
8 PAFZZ PAOZZ 5940-01-110-6423 30554 88-20274-5 ...TERMINAL, SPADE #6, 16-14 AWG 6
9 PAFZZ PAOZZ 5940-00-113-8184 96906 MS25036-150 ...TERMINAL, LUG 1/4, 22-18 AWG 5
10 PAFZZ PAOZZ 5940-00-143-4794 96906 MS25036-112 ...TERMINAL, LUG #10, 12-10 AWG 1
11 PAFZZ PAOZZ 5940-01-112-9746 30554 88-20275-1 ...TERMINAL, QUICKDISC 4
12 PAFZZ PAOZZ 5940-00-143-4771 96906 MS25036-103 ...TERMINAL, LUG .... 6 13 PAFZZ PAOZZ 5940-00-115-0763 96906 MS20659-140 ...TERMINAL, LUG
UOC: EMK, YNN 1
13 PAFZZ PAOZZ 5940-00-283-5281 96906 MS25036-109 ...TERMINAL, LUG
UOC: 86Q, 86R............................ 1
14 PAFZZ PAOZZ 5940-00-114-1306 96906 MS25036-117 ...TERMINAL, LUG 3
15 PAFZZ PAOZZ 5940-00-682-2445 96906 MS25036-158 ...TERMINAL, LUG 2
16 PAFZZ PAOZZ 5940-00-813-0698 96906 MS25036-101 ...TERMINAL, LUG 1
17 PAFZZ PAOZZ 5940-00-143-4774 96906 MS25036-153 ...TERMINAL, LUG 1
18 PAFZZ PAOZZ 5940-00-504-4703 96906 MS25036-104 ...TERMINAL, LUG 1
19 PAFZZ PAOZZ 5940-01-369-6948 30554 88-22119-12 ...TERMINAL, LUG 1
20 PAFZZ PAOZZ 5940-00-113-8183 96906 MS25036-113 ...TERMINAL, LUG 1
21 PAFZZ PAOZZ 5940-00-143-4780 96906 MS25036-108 ...TERMINAL, LUG 1
22 PAFZZ MOO 6145-00-284-0657 81349 M5086/2-8-9 . . . WIRE, ELECTRIC 1
23 PAFZZ MOO 6145-00-578-7514 81349 M5086/2-12-9 . . . WIRE, ELECTRIC 1
24 PAFZZ MOO 6145-00-578-7513 81349 M5086/2-10-9 . . . WIRE, ELECTRIC 1
25 PAFZZ PAOZZ 5975-00-111-3208 96906 MS3367-5-9 ...STRAP, TIEDOWN,ELECT V
26 PAFZZ PAOZZ 5975-00-074-2072 96906 MS3367-1-9 ...STRAP, TIEDOWN,ELECT V
27 PAFZZ PAOZZ 5975-00-944-1499 96906 MS3368-1-9A ...STRAP, TIEDOWN,ELECT 1
28 PAFZZ PAOZZ 5310-00-836-3520 30554 69-561-1 . . . NUT, PLAIN, ASSEMBLED 4

(1) (2) (3) (4) (5) (6) (7)
SMR CODE
a. b.
ITEM NO ARMY AIR FORCE NSN CAGEC PART NUMBER DESCRIPTION AND USABLE ON CODE (UOC) QTY

29 PAFZZ PAOZZ 5305-01-367-2314 30554 69-662-11 ...SCREW, ASSEMBLED WAS 4
30 PAFZZ PAOZZ 5935-01-014-4920 30554 88-21947 ...ADAPTER,CABLECLAMP 2
31 PAFZZ PAOZZ 5999-01-170-0558 30554 88-21942 ...CONTACT,ELECTRICAL 38
32 PAFZA PAOZA 5999-01-366-7952 30554 88-21944 ...CONTACT,ELECTRICAL 7
33 PAFZA PAOZA 5935-01-101-7828 30554 88-21982 ...CONNECTOR BODY,PLUG 1
34 PAFZZ PAOZZ 5935-01-023-5445 30554 88-21940 ...CONNECTOR BODY,PLUG 1
35 PAFZZ MOO 6145-00-851-8505 81349 M5086/2-20-9 . . . WIRE, ELECTRIC 1
36 XDFZZ XB 5940-01-470-2470 30554 88-21181-2 ...TERMINALBLOCK TB8 1
37 PAFZZ MOO 6145-00-578-6605 81349 M5086/2-16-9 . . . WIRE, ELECTRIC 1
38 XDFZZ XB 5999-01-366-2621 30554 88-21183-2 ...TERMINALBLOCK,JUMPER 1
39 XDFZZ XB 5940-01-478-1154 30554 88-21181-10 ...TERMINALBLOCK TB7 1
40 PAFZZ PAOZZ 5999-01-039-8438 30554 88-20477 ...CONTACT,ELECTRICAL 11
41 PAFZZ PAOZZ 5935-00-315-9563 30554 88-20474 ...CONNECTOR BODY,RECE 1
42 PAFZZ PAOZZ 5935-01-012-1273 30554 88-20470 ...CONNECTOR BODY,RECE .... 2 42 PAFZZ PAOZZ 5935-01-012-1273 30554 88-20470 ...CONNECTOR BODY,RECE
UOC: 86Q, 86R............................ 1
43 AFFFF AOO 30554 88-22101-10 ...CABLE ASSY, SHIELDED .... 1 44 MFFZZ MOO 30554 88-22101-10-1 ....CABLE, SHIELDED MAKE FROM
P/N M27500-16TE2T14 (81349), 59 INCHES REQUIRED 1
45 PAFZZ PAOZZ 5940-01-531-5519 30554 88-20596-1 ....SHIELD, TERMINATION CUT
WIRE LEAD TO 6.5 INCHES 1
46 PAFZA PAOZA 5999-01-366-7952 30554 88-21944 ....SOCKET 2
47 PAFZZ PAOZZ 5940-01-369-6948 30554 88-22119-12 ....TERMINAL, LUG 1
48 PAFZZ PAOZZ 5940-00-283-5281 96906 MS25036-109 ....TERMINAL, LUG 1
49 PAFZZ PAOZZ 5940-00-113-8185 96906 MS25036-151 ....TERMINAL, LUG .... 1 50 MFFZZ MOO 30554 88-22629-043 ....MARKER,IDENTIFICATION
MAKE FROM P/N M23053/5-105-9 (81349), 1.25 INCHES REQUIRED 2
51 MFFZZ MOO 30554 88-22629-044 ....MARKER,IDENTIFICATION
MAKE FROM P/N M23053/5-105-9 (81349), 1.25 INCHES REQUIRED 2
52 MFFZZ MOO 30554 88-22629-050 ....MARKER,IDENTIFICATION
MAKE FROM P/N M23053/5-105-9 (81349), 1.25 INCHES REQUIRED 1
53 AFFFF AOO 30554 88-22101-11 ...CABLE ASSY, SHIELDED (NOT
SHOWN) 1
54 MFFZZ MOO 30554 88-22101-11-1 ....CABLE, SHIELDED MAKE FROM
P/N M27500-16TE2T14 (81349), 76 INCHES REQUIRED 1
55 PAFZZ PAOZZ 5940-01-531-5519 30554 88-20596-1 ....SHIELD, TERMINATION CUT
WIRE LEAD TO 6.5 INCHES 1
56 PAFZA PAOZA 5999-01-366-7952 30554 88-21944 ....SOCKET 2
57 PAFZZ PAOZZ 5940-00-143-4780 96906 MS25036-108 ....TERMINAL, LUG 1
58 PAFZZ PAOZZ 5940-00-143-4771 96906 MS25036-103 ....TERMINAL, LUG 1
59 PAFZZ PAOZZ 5940-01-082-3321 30554 88-20275-3 ....TERMINAL, QUICKDISC .... 1 60 MFFZZ MOO 30554 88-22629-046 ....MARKER,IDENTIFICATION
MAKE FROM P/N M23053/5-105-9 (81349), 1.25 INCHES REQUIRED 2
61 MFFZZ MOO 30554 88-22629-051 ....MARKER,IDENTIFICATION
MAKE FROM P/N M23053/5-105-9 (81349), 1.25 INCHES REQUIRED 2
62 MFFZZ MOO 30554 88-22629-052 ....MARKER,IDENTIFICATION
MAKE FROM P/N M23053/5-105-9 (81349), 1.25 INCHES REQUIRED 2

(1) ITEM NO	(2) SMR CODE a. ARMY	b. AIR FORCE	(3) NSN	(4) CAGEC	(5) PART NUMBER	(6) DESCRIPTION AND USABLE ON CODE (UOC)	(7) QTY

63 AFFFF AOO 30554 88-22103-7 ...CABLE ASSY, SHIELDED (NOT
SHOWN) 1
64 MFFZZ MOO 30554 88-22103-7-1 ....CABLE, SHIELDED MAKE FROM
P/N M27500-16TE2T14 (81349), 55
INCHES REQUIRED 1
65 PAFZZ PAOZZ 5940-01-531-5519 30554 88-20596-1 ....SHIELD, TERMINATION CUT
WIRE LEAD TO 7 INCHES 1
66 PAFZA PAOZA 5999-01-366-7952 30554 88-21944 ....SOCKET 2
67 PAFZZ PAOZZ 5999-01-170-0558 30554 88-21942 ....CONTACT,ELECTRICAL 1
68 PAFZZ PAOZZ 5940-00-283-5280 96906 MS25036-106 ....TERMINAL, LUG .... 2 69 MFFZZ MOO 30554 88-22629-063 ....MARKER,IDENTIFICATION
MAKE FROM P/N M23053/5-105-9 1.25
INCHES LONG 2
70 MFFZZ MOO 30554 88-22629-064 ....MARKER,IDENTIFICATION
MAKE FROM P/N M23053/5-105-9
(81349), 1.25 INCHES REQUIRED 2
71 MFFZZ MOO 30554 88-22629-074 ....MARKER,IDENTIFICATION
MAKE FROM P/N M23053/5-105-9
(81349), 1.25 INCHES REQUIRED 1
72 AFFFF AOO 30554 88-22103-8 ...CABLE ASSY, SHIELDED (NOT
SHOWN) 1
73 MFFZZ MOO 30554 88-22103-8-1 ....CABLE, SHIELDED MAKE FROM
P/N M27500-16TE2T14 (81349), 49
INCHES REQUIRED 1
74 PAFZZ PAOZZ 5940-01-531-5519 30554 88-20596-1 ....SHIELD, TERMINATION CUT
WIRE LEAD TO 7 INCHES 1
75 PAFZA PAOZA 5999-01-366-7952 30554 88-21944 ....SOCKET 2
76 PAFZZ PAOZZ 5999-01-170-0558 30554 88-21942 ....CONTACT,ELECTRICAL 1
77 PAFZZ PAOZZ 5940-01-082-3321 30554 88-20275-3 ....TERMINAL, QUICKDISC 1
78 PAFZZ PAOZZ 5940-01-139-0853 30554 88-20275-4 ....TERMINAL, QUICKDISC .... 1 79 MFFZZ MOO 30554 88-22629-066 ....MARKER,IDENTIFICATION
MAKE FROM P/N M23053/5-105-9
(81349), 1.25 INCHES REQUIRED 2
80 MFFZZ MOO 30554 88-22629-067 ....MARKER,IDENTIFICATION
MAKE FROM P/N M23053/5-105-9
(81349), 1.25 INCHES REQUIRED 2
81 MFFZZ MOO 30554 88-22629-075 ....MARKER,IDENTIFICATION
MAKE FROM P/N M23053/5-105-9
(81349), 1.25 INCHES REQUIRED 1
82 AFFFF AOO 30554 88-22103-9 ...CABLE ASSY, SHIELDED (NOT
SHOWN) 1
83 MFFZZ MOO 30554 88-22103-9-1 ....CABLE, SHIELDED MAKE FROM
P/N M27500-16TE2T14 (81349), 50
INCHES REQUIRED 1
84 PAFZZ PAOZZ 5940-01-531-5519 30554 88-20596-1 ....SHIELD, TERMINATION CUT
WIRE LEAD TO 7 INCHES 1
85 PAFZA PAOZA 5999-01-366-7952 30554 88-21944 ....SOCKET 2
86 PAFZZ PAOZZ 5999-01-170-0558 30554 88-21942 ....CONTACT,ELECTRICAL 1
87 PAFZZ PAOZZ 5999-01-039-8438 30554 88-20477 ....CONTACT,ELECTRICAL .... 2 88 MFFZZ MOO 30554 88-22629-069 ....MARKER,IDENTIFICATION
MAKE FROM P/N M23053/5-105-9
(81349), 1.25 INCHES REQUIRED 2
89 MFFZZ MOO 30554 88-22629-070 ....MARKER,IDENTIFICATION
MAKE FROM P/N M23053/5-105-9
(81349), 1.25 INCHES REQUIRED 2

(1)	(2)		(3)	(4)	(5)	(6)	(7)
	SMR CODE						
	a.	b.					
ITEM NO	ARMY	AIR FORCE	NSN	CAGEC	PART NUMBER	DESCRIPTION AND USABLE ON CODE (UOC)	QTY
90	MFFZZ	MOO		30554	88-22629-076	....MARKER,IDENTIFICATION MAKE FROM P/N M23053/5-105-9 (81349), 1.25 INCHES REQUIRED	1
91	PAFZA	PAFZA	5935-00-852-9611	96906	MS3102R18-11P	...CONNECTOR,RECEPTACL J16	1
92	PAFZZ	PAOZZ	5935-01-020-4094	30554	88-20472	...CONNECTOR BODY,RECE UOC: 86Q, 86R.............................	1

END OF FIGURE

FIELD AND SUSTAINMENT MAINTENANCE

15 kW 50/60 AND 400 Hz SKID MOUNTED TACTICAL QUIET GENERATOR SETS

GROUP 07 OUTPUT BOX ASSEMBLY

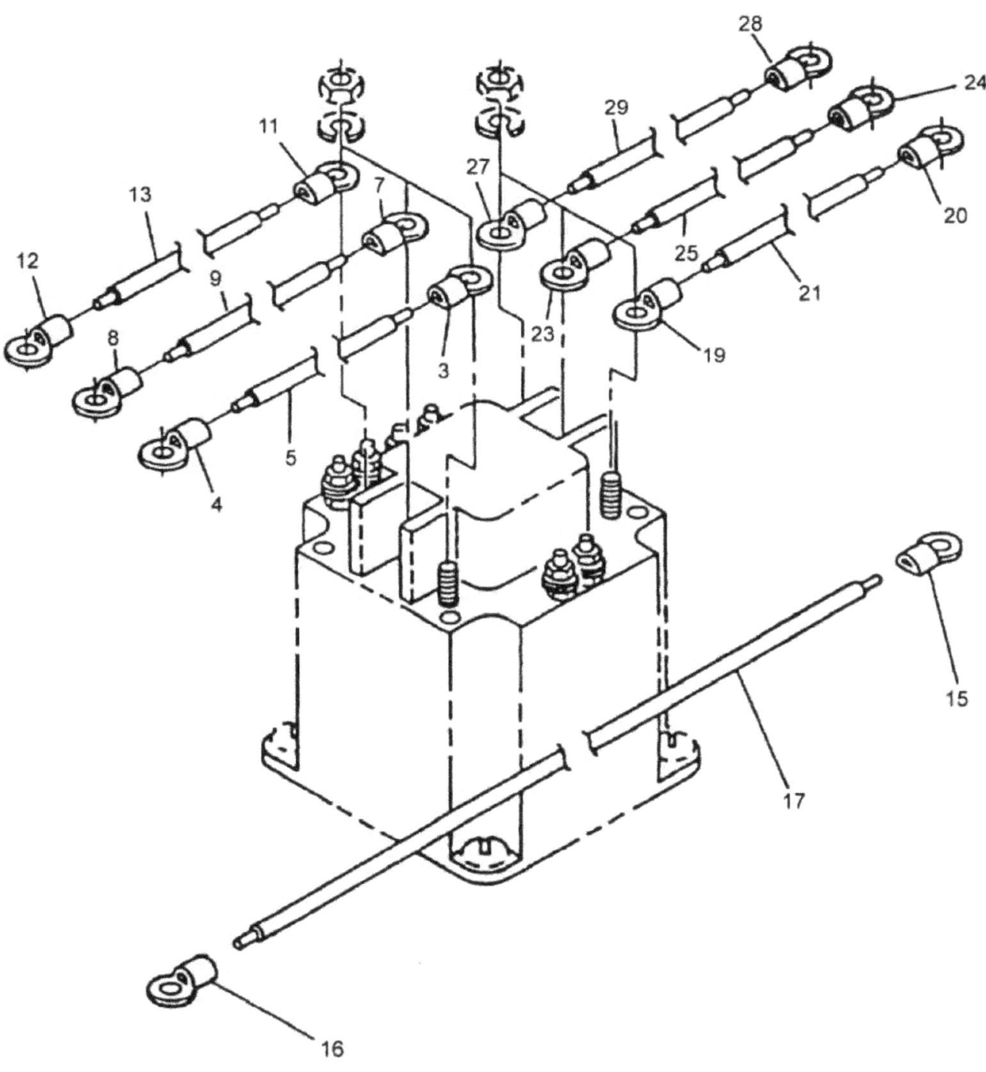

Figure 19. Output Box Assembly (Sheet 1 of 2).

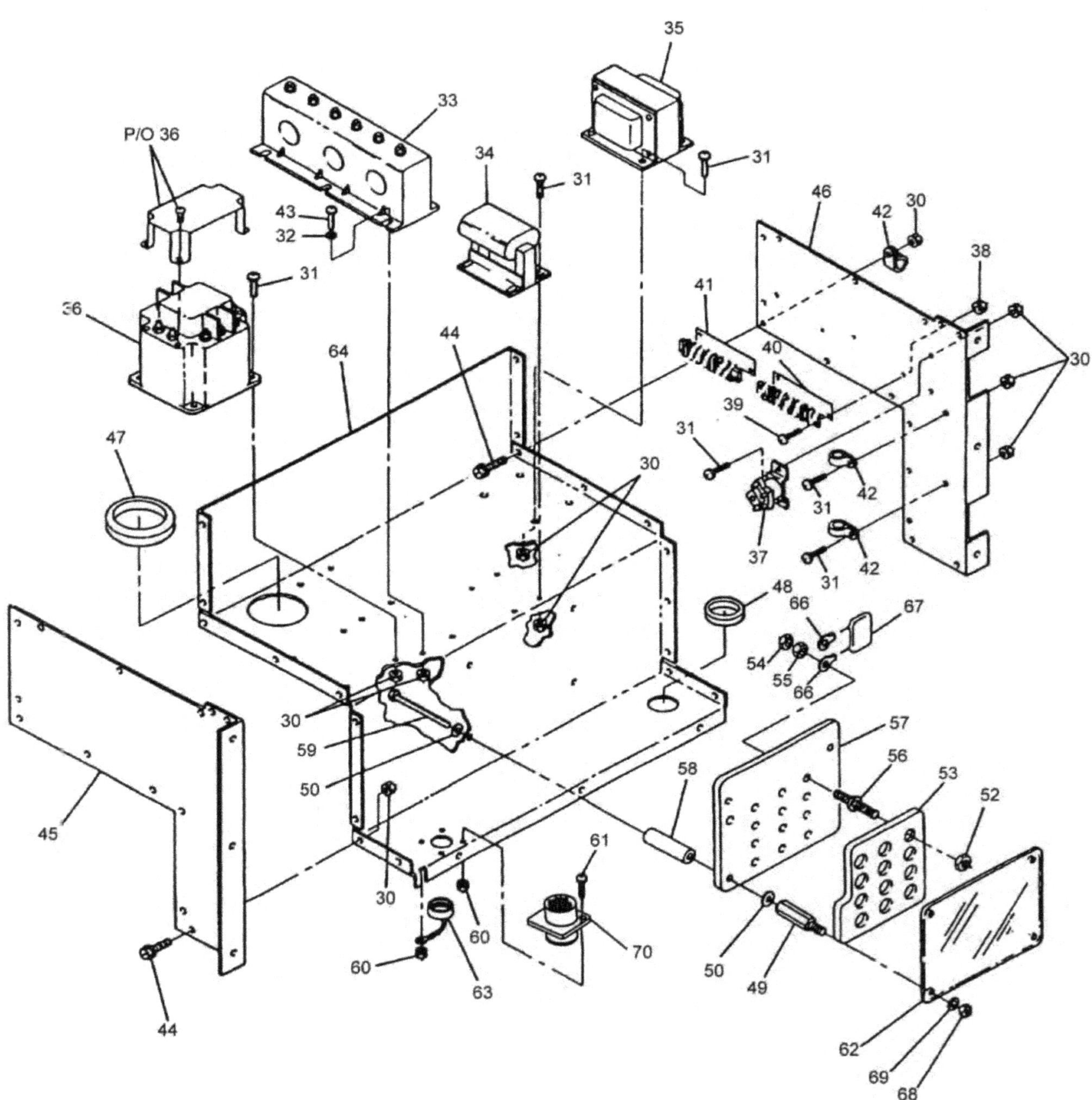

Figure 19. Output Box Assembly (Sheet 2 of 2).

(1)	(2) SMR CODE a.	b.	(3)	(4)	(5)	(6)	(7)
ITEM NO	ARMY	AIR FORCE	NSN	CAGEC	PART NUMBER	DESCRIPTION AND USABLE ON CODE (UOC)	QTY

GROUP 07 OUTPUT BOX ASSEMBLY

FIG. 19 OUTPUT BOX ASSEMBLY

1 XDFFF XB 30554 88-21580 . OUTPUT BOX ASSY 60 HZ (NOT SHOWN)
UOC: EMK 1

1 XDFFF XB 30554 88-21582 . OUTPUT BOX ASSY 400 HZ (NOT SHOWN)
UOC: YNN 1

1 XDFFF XB 30554 97-24082 . OUTPUT BOX ASSY 50/60 HZ AND 400 HZ (NOT SHOWN)
UOC: 86Q, 86R............................ 1

2 AFFFF AOO 30554 88-22126-1 . . CABLE ASSY, AC POWER (NOT SHOWN) 1

3 PAFZZ PAOZZ 5940-01-369-2271 30554 88-22119-3 ...TERMINAL, LUG 1

4 PAFZZ PAOZZ 5940-00-113-8191 96906 MS25036-127 ...TERMINAL, LUG .... 1 5 MFFZZ MOO 30554 88-22126-1-1 . . . WIRE, ELECTRIC MAKE FROM P/N M5086/2-2-9 (81349), 14.5 INCHES REQUIRED 1

6 AFFFF AOO 30554 88-22126-2 . . CABLE ASSY, AC POWER (NOT SHOWN) 1

7 PAFZZ PAOZZ 5940-01-369-2271 30554 88-22119-3 ...TERMINAL, LUG 1

8 PAFZZ PAOZZ 5940-00-113-8191 96906 MS25036-127 ...TERMINAL, LUG .... 1 9 MFFZZ MOO 30554 88-22126-2-1 . . . WIRE, ELECTRIC MAKE FROM P/N M5086/2-2-9 (81349), 17.0 INCHES REQUIRED 1

10 AFFFF AOO 30554 88-22126-3 . . CABLE ASSY, AC POWE (NOT SHOWN) 1

11 PAFZZ PAOZZ 5940-01-369-2271 30554 88-22119-3 ...TERMINAL, LUG 1

12 PAFZZ PAOZZ 5940-00-113-8191 96906 MS25036-127 ...TERMINAL, LUG .... 1 13 MFFZZ MOO 30554 88-22126-3-1 . . . WIRE, ELECTRIC MAKE FROM P/N M5086/2-2-9 (81349), 20.0 INCHES REQUIRED 1

14 AFFFF AOO 30554 88-22126-4 . . CABLE ASSY, AC POWER (NOT SHOWN) 1

15 PAFZZ PAOZZ 5940-00-113-8191 96906 MS25036-127 ...TERMINAL, LUG 1

16 PAFZZ PAOZZ 5940-00-113-9831 96906 MS25036-128 ...TERMINAL, LUG .... 1 17 MFFZZ MOO 30554 88-22126-4-1 . . . WIRE, ELECTRIC MAKE FROM P/N M5086/2-2-9 (81349), 41.0 INCHES REQUIRED 1

18 AFFFF AOO 30554 88-22126-5 . . CABLE ASSY, AC POWER (NOT SHOWN) 1

19 PAFZZ PAOZZ 5940-01-369-2271 30554 88-22119-3 ...TERMINAL, LUG 1

20 PAFZZ PAOZZ 5940-00-113-9831 96906 MS25036-128 ...TERMINAL, LUG .... 1 21 MFFZZ MOO 30554 88-22126-5-1 . . . WIRE, ELECTRIC MAKE FROM P/N M5086/2-2-9 (81349), 31 INCHES REQUIRED 1

22 AFFFF AOO 30554 88-22126-6 . . CABLE ASSY, AC POWER (NOT SHOWN) 1

23 PAFZZ PAOZZ 5940-01-369-2271 30554 88-22119-3 ...TERMINAL, LUG 1

24 PAFZZ PAOZZ 5940-00-113-9831 96906 MS25036-128 ...TERMINAL, LUG 1

(1) (2) (3) (4) (5) (6) (7)
SMR CODE
a. b.
ITEM NO ARMY AIR FORCE NSN CAGEC PART NUMBER DESCRIPTION AND USABLE ON CODE (UOC) QTY

25 MFFZZ MOO 30554 88-22126-6-1 . . . WIRE, ELECTRIC MAKE FROM P/
N M5086/2-2-9 (81349), 33.0 INCHES
REQUIRED 1
26 AFFFF AOO 30554 88-22126-7 . . CABLE ASSY, AC POWER (NOT
SHOWN) 1
27 PAFZZ PAOZZ 5940-01-369-2271 30554 88-22119-3 ...TERMINAL, LUG 1
28 PAFZZ PAOZZ 5940-00-113-9831 96906 MS25036-128 ...TERMINAL, LUG .... 1 29 MFFZZ MOO 30554 88-22126-7-1 . . . WIRE, ELECTRIC MAKE FROM P/
N M5086/2-2-9 (81349), 35.0 INCHES
REQUIRED ....
30 PAFZZ PAOZZ 5310-01-012-3595 30554 69-561-6 . . NUT, PLAIN, ASSEMBLED 40
31 PAFZZ PAOZZ 5305-00-191-6226 30554 69-662-64 . . SCREW, ASSEMBLED WAS 22
32 PAFZZ PAOZZ 5310-01-234-9416 96906 MS51412-2 . . WASHER, FLAT 6
33 XDFZZ XB 5950-01-368-2915 30554 88-21131 . . TRANSFORMER, CURRENT 1
34 PAFZZ PAOZZ 5950-01-382-3371 30554 88-21015 . . TRANSFORMER, CURRENT .... 1 35 PAFZZ PAOZZ 5950-01-368-3006 30554 01-21504-2 . . TRANSFORMER, POWER,
UOC: YNN, 86R............................ 1
35 PAFZZ PAOZZ 5950-01-370-3328 30554 01-21504-1 . . TRANSFORMER, POWER,
UOC: EMK, 86Q............................ 1
36 PAFZZ PAOZZ 6110-01-367-8921 30544 88-21066 . . CONTACTOR, MAGNETIC 1
37 PAFZZ PAOZZ 5945-01-365-9953 30554 88-21082 . . RELAY, ELECTROMAGNET 1
38 PAFZZ PAOZZ 5310-00-052-3632 30554 69-561-3 . . NUT, PLAIN, ASSEMBLED 4
39 PAFZZ PAOZZ 5305-00-036-6902 30554 69-662-36 . . SCREW, ASSEMBLED WAS 4
40 XDFZZ XB 5940-01-470-3031 30554 88-21182-10 . . MARKER, STRIP 1
41 XDFZZ XB 5940-01-470-2768 30554 88-21182-2 . . MARKER, STRIP 1
42 PAFZZ PAOZZ 5340-00-724-7038 96906 MS21333-76 . . CLAMP, LOOP 3
43 PAFZZ PAOZZ 5305-01-187-5878 78189 69-662-65 . . SCREW, ASSEMBLED 6
44 PAFZZ PAOZZ 5306-01-156-7663 30554 88-20260-21 . SCREW, HEX WASHERHEAD 18
45 XDFZZ XB 30554 88-21932 . . PANEL, OUTPUT BOX, LS 1
46 XDFZZ XB 30554 88-22060 . . PANEL, OUTPUT BOX, RS 1
47 PAFZZ PAOZZ 5325-00-351-4543 30554 88-22122 . . GROMMET, NONMETALLIC 1
48 PAFZZ PAOZZ 5325-00-290-1960 96906 MS35489-27 . . GROMMET, NONMETALLIC 1
49 PAFZZ PAOZZ 5340-01-222-4225 30554 72-2098-2 . . POST, ELECTRICAL-MEC 4
50 PAFZZ PAOZZ 5310-01-103-6042 96906 MS51412-4 . . WASHER, FLAT .... 8 51 PAFZZ PAOZZ 5940-01-384-0384 30554 88-22437 . . TERMINAL BOARD, ASSY (NOT
SHOWN) 1
52 PAFZZ PAOZZ 5310-01-466-7247 30554 88-22786 . . . NUT, SELF-LOCKING, HEX 13
53 PAFZZ PAOZZ 5940-01-369-2877 30554 88-22436 ...TERMINALBOARD 1
54 PAFZZ PAOZZ 5310-01-366-4412 30554 88-21930-1 . . . NUT, SELF-LOCKING, HEX 13
55 PAFZZ PAOZZ 5310-01-289-7716 30554 72-2062-1 . . . NUT, PLAIN, HEX 13
56 PAFZZ PAOZZ 5940-01-003-8579 30554 72-2236 ...TERMINAL, STUD 13
57 PAFZZ PAOZZ 5940-00-606-4970 30554 72-2213 ...TERMINALBOARD 1
58 XDFZZ XB 30554 88-21669 . . MOUNT, RECONNECTION 1
59 PAFZZ PAOZZ 5940-00-606-4970 96906 MS51493-2 . . SCREW, MACHINE 4
60 PAFZZ PAOZZ 5940-00-606-4970 30554 69-561-1 . . NUT, PLAIN, ASSEMBLED 4
61 PAFZZ PAOZZ 5940-00-606-4970 30554 69-662-5 . . SCREW, ASSEMBLED WASH 4
62 PAFZZ PAOZZ 5940-00-606-4970 30554 88-22509 . . COVER, PROTECTIVE 1
63 PAFZZ PAOZZ 5940-00-606-4970 30554 88-20548-1 . . COVER, ELECTRICAL CON 1
64 XDFZZ XB 30554 88-22039 . . PANEL, OUTPUT BOX, REAR .... 1 65 AFFFF AOO 5910-01-384-1745 30554 88-22758 . . CAPACITOR ASSY, EMI (NOT
SHOWN) 3
66 PAFZZ PAOZZ 5940-00-113-3140 96906 MS20659-125 ...TERMINAL, LUG 2
67 PAFZZ PAOZZ 5910-01-343-8827 30554 88-22757-2 ...CAPACITOR 1

(1) ITEM NO	(2) SMR CODE a. ARMY	(2) SMR CODE b. AIR FORCE	(3) NSN	(4) CAGEC	(5) PART NUMBER	(6) DESCRIPTION AND USABLE ON CODE (UOC)	(7) QTY
68	PAFZZ	PAOZZ	5310-00-929-1807	96906	MS51922-2	. . NUT, SELF-LOCKING, HEX	4
69	PAFZZ	PAOZZ	5310-01-103-6042	96906	MS51412-4	. . WASHER, FLAT	4
70	PBFFF	PBFFF	6150-01-407-8102	30554	88-22164	. . WIRING HARNESS, BR (SEE FIGURE 18 FOR PARTS BREAKDOWN) UOC: EMK, YNN	REF
70	PBFFF	PBFFF		30554	97-24048	. . WIRING HARNESS, BR (SEE FIGURE 18 FOR PARTS BREAKDOWN) UOC: 86Q, 86R	REF

END OF FIGURE

FIELD AND SUSTAINMENT MAINTENANCE

15 kW 50/60 AND 400 Hz SKID MOUNTED TACTICAL QUIET GENERATOR SETS
GROUP 08 OUTPUT LOAD TERMINAL BOARD: LOAD TERMINAL BOARD

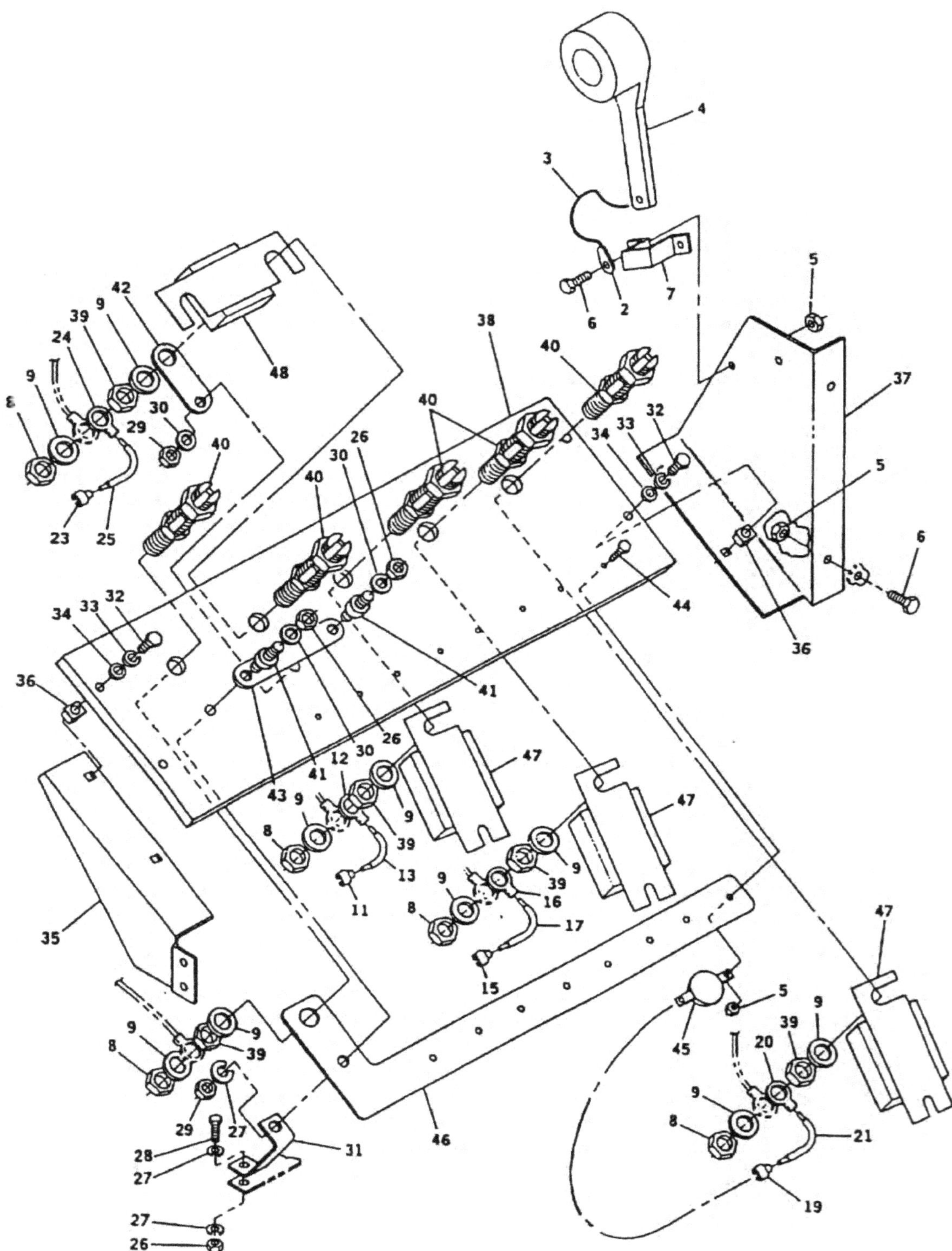

Figure 20. Load Terminal Board.

(1)	(2)		(3)	(4)	(5)	(6)	(7)
	SMR CODE						
	a.	b.					
ITEM NO	ARMY	AIR FORCE	NSN	CAGEC	PART NUMBER	DESCRIPTION AND USABLE ON CODE (UOC)	QTY

GROUP 08 OUTPUT LOAD TERMINAL BOARD

FIG. 20 LOAD TERMINAL BOARD

1 AFFFF AOO 4020-01-470-6597 30554 88-22469 . CORD, LOAD, WRENCH (NOT SHOWN) 1

2 PAFZZ PAOZZ 5940-00-115-2677 96906 MS20659-144 . . TERMINAL, LUG #4 AWG, #10 STUD SIZE 1

3 MFFZZ MOO 30554 88-22469-2 . . ROPE, FIBROUS MAKE FROM P/N 1-41NDIA (72205), 30 INCHES REQUIRED 1

4 PAFZZ PAOZZ 5120-01-373-8976 30554 88-21146 . WRENCH, BOX 1

5 PAFZZ PAOZZ 5310-01-012-3595 30554 69-561-6 . NUT, PLAIN, ASSEMBLED 12

6 PAFZZ PAOZZ 5306-01-156-7663 30554 88-20260-21 . SCREW, HEX WASHERHEAD 12

7 XDFZZ XB 30554 88-22136 . BRACKET, WRENCH 1

8 PAFZZ PAOZZ 5310-01-183-5529 30554 88-21930-2 . NUT, SELF LOCKING 5

9 PAFZZ PAOZZ 5310-00-465-2719 30554 88-20033-40B . WASHER, FLAT .... 10 10 AFFFF AOO 30554 88-20305-1 . WIRE, VARISTOR (L1) (NOT SHOWN) 1

11 PAFZZ PAOZZ 5940-01-082-3321 30554 88-20275-3 . . TERMINAL, QUICK DISC 1

12 PAFZZ PAOZZ 5940-01-369-6948 30554 88-22119-12 . . TERMINAL, LUG .... 1 13 MFFZZ MOO 30554 88-20305-1-1 . . WIRE, ELECTRIC MAKE FROM P/N M5086/2-16-9 (81349), 6.0 INCHES REQUIRED 1

14 AFFFF AOO 30554 88-20305-2 . WIRE, VARISTOR (L2) (NOT SHOWN) 1

15 PAFZZ PAOZZ 5940-01-082-3321 30554 88-20275-3 . . TERMINAL, QUICK DISC 1

16 PAFZZ PAOZZ 5940-01-369-6948 30554 88-22119-12 . . TERMINAL, LUG .... 1 17 MFFZZ MOO 30554 88-20305-2-1 . . WIRE, ELECTRIC MAKE FROM P/N M5086/2-16-9 (81349), 6.0 INCHES REQUIRED 1

18 AFFFF AOO 30554 88-20305-3 . WIRE, VARISTOR (L3) (NOT SHOWN) 1

19 PAFZZ PAOZZ 5940-01-082-3321 30554 88-20275-3 . . TERMINAL, QUICK DISC 1

20 PAFZZ PAOZZ 5940-01-369-6948 30554 88-22119-12 . . TERMINAL, LUG .... 1 21 MFFZZ MOO 30554 88-20305-3-1 . . WIRE, ELECTRIC MAKE FROM P/N M5086/2-16-9 (81349), 6.0 INCHES REQUIRED 1

22 AFFFF AOO 5905-01-463-0058 30554 88-20305-5 . WIRE, VARISTOR (LO) (NOT SHOWN) 10

23 PAFZZ PAOZZ 5940-01-082-3321 30554 88-20275-3 . . TERMINAL, QUICK DISC 1

24 PAFZZ PAOZZ 5940-01-369-6948 30554 88-22119-12 . . TERMINAL, LUG .... 1 25 MFFZZ MOO 30554 88-20305-5-1 . . WIRE, ELECTRIC MAKE FROM P/N M5086/2-16-9 (81349). 6.0 INCHES REQUIRED 1

26 PAFZZ PAOZZ 5310-01-466-7247 30554 88-22786 . NUT, SELF-LOCKING, HEX 3

27 PAFZZ PAOZZ 5310-01-466-7321 30554 88-22792-2 . WASHER, LOCK 3

28 PAFZZ PAOZZ 5305-00-543-4372 80204 B1821BH038C075N . SCREW, CAP, HEX HEAD 1

29 PAFZZ PAOZZ 5310-01-366-4412 30554 88-21930-1 . NUT, SELF-LOCKING, HEX 2

30 PAFZZ PAOZZ 5310-00-205-9951 30554 88-20033-26B . WASHER, FLAT 3

31 XDFZZ XB 30554 88-22459 . STRAP, GROUND 1

32 PAFZZ PAOZZ 5305-00-068-0509 80204 B1821BH025C125N . SCREW, CAP, HEX HEAD 4

(1) (2) (3) (4) (5) (6) (7)
SMR CODE
a. b.
ITEM NO ARMY AIR FORCE NSN CAGEC PART NUMBER DESCRIPTION AND USABLE ON CODE (UOC) QTY

33 PAFZZ PAOZZ 5310-00-274-8715 80205 MS35338-63 . WASHER, LOCK 4
34 PAFZZ PAOZZ 5310-01-103-6042 96906 MS51412-4 . WASHER, FLAT 4
35 XDFZZ XB 30554 88-21767 . SUPPORT, LOAD BOARD 1
36 PAFZZ PAOZZ 5340-00-297-0312 30554 88-21674-3 . NUT, CAGE 4
37 XDFZZ XB 30554 88-21933 . SUPPORT, LOAD BOARD 1
38 XDFZZ XB 30554 88-21666 . BOARD, LOAD TERMINAL 1
39 PAFZZ PAOZZ 5310-00-189-8467 30554 88-22336-1 . NUT, PLAIN, HEX 5
40 PAFZZ PAOZZ 5940-00-958-1214 30554 69-692-1 . TERMINAL, STUD 5
41 PAFZZ PAOZZ 5940-01-003-8579 30554 72-2236 . TERMINAL, STUD 5
42 XDFZZ XB 30554 88-22163-1 . BAR, BUS GROUND 1
43 XDFZZ XB 30554 88-22146 . BUS, BAR 1
44 PAFZZ PAOZZ 5305-01-365-6314 30554 88-20260-25 . SCREW, CAP, HEX HEAD 8
45 PAFZZ PAOZZ 5905-01-063-9644 30554 88-21179 . RESISTOR, VOLTAGE SE 4
46 XDFZZ XB 30554 88-21775 . BAR, GROUND PLANE .... 1 47 PAFZZ PAOZZ 5915-01-396-9253 30554 88-22761 . FILTER ASSEMBLY, ELE L1, L2, AND
L3
UOC: EMK, 86Q............................ 3
47 PAFZZ PAOZZ 5915-01-488-5004 30554 88-22805 . FILTER ASSEMBLY, ELE L1, L2, AND
L3
UOC: YNN, 86R............................ 3
48 PAFZZ PAOZZ 5915-01-394-0942 30554 88-22762 . FILTER ASSEMBLY, ELE LO AND
GND
UOC: EMK, 86Q............................ 1
48 PAFZZ PAOZZ 5915-01-466-8148 30554 88-22807 . FILTER ASSEMBLY, ELE LO AND
GND
UOC: YNN, 86R............................ 1

END OF FIGURE

FIELD AND SUSTAINMENT MAINTENANCE

15 kW 50/60 AND 400 Hz SKID MOUNTED TACTICAL QUIET GENERATOR SETS

GROUP 09 ENGINE ACCESSORIES

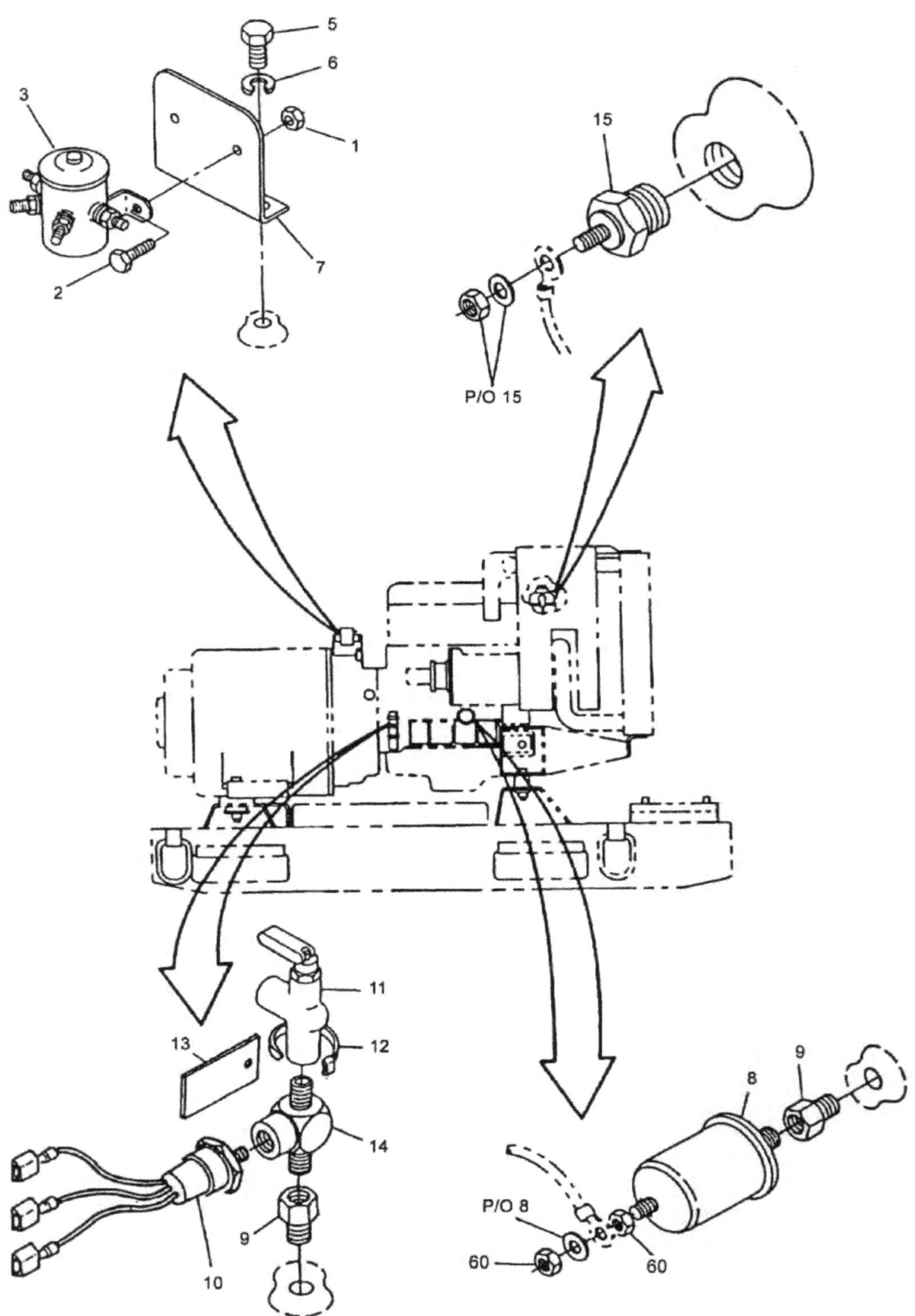

Figure 21. Engine Accessories (Sheet 1 of 5).

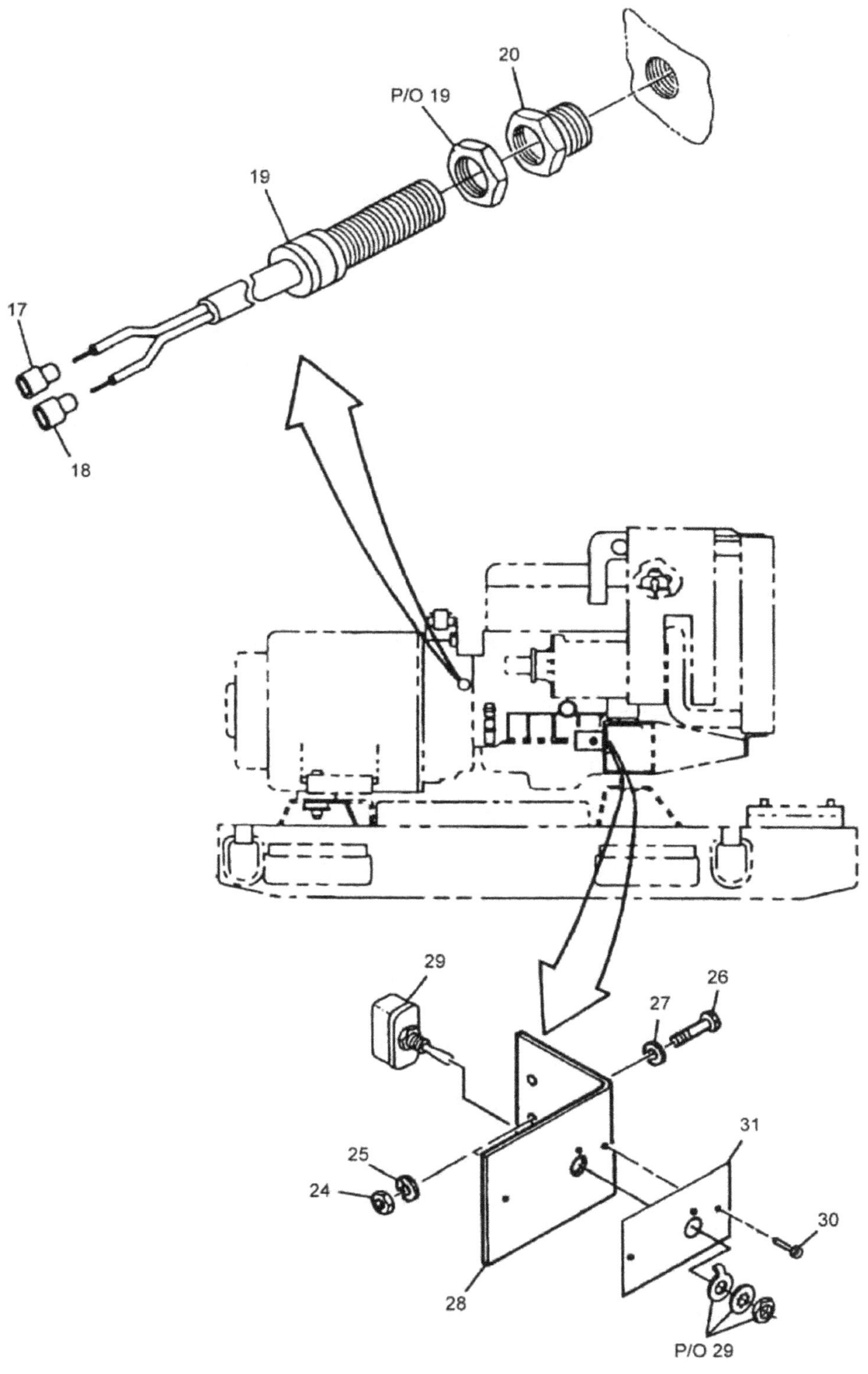

Figure 21. Engine Accessories (Sheet 2 of 5).

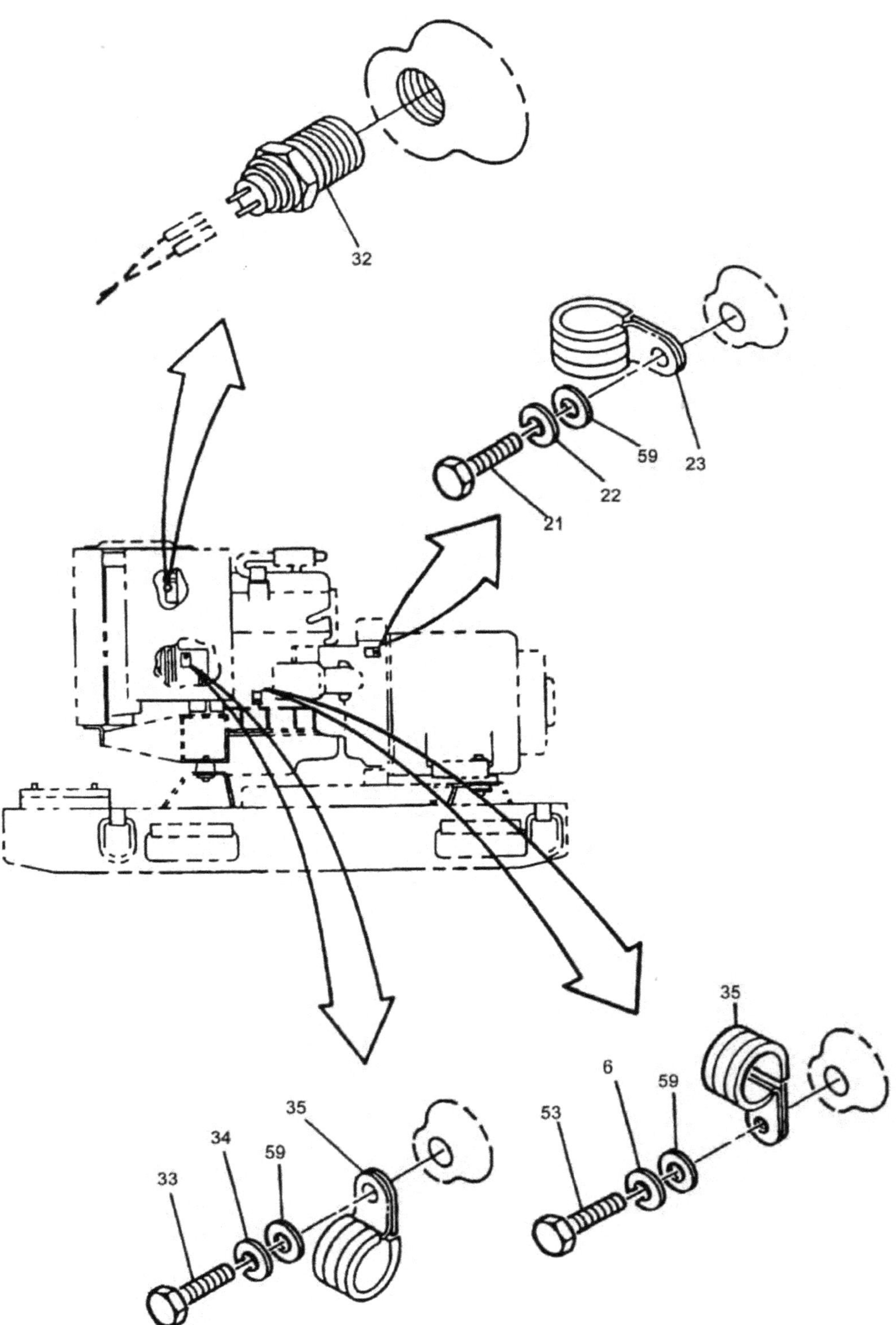

Figure 21. Engine Accessories (Sheet 3 of 5).

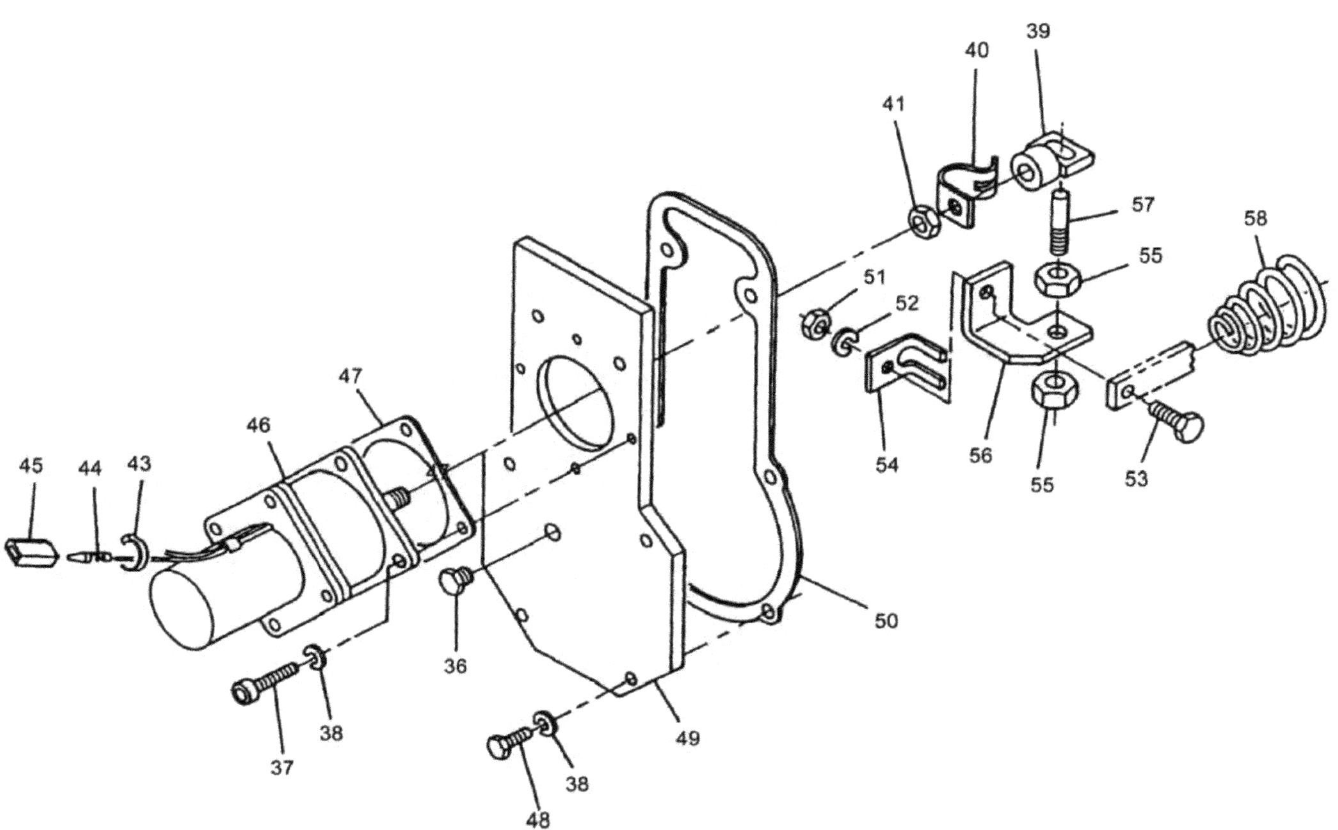

Figure 21. Engine Accessories (Sheet 4 of 5).

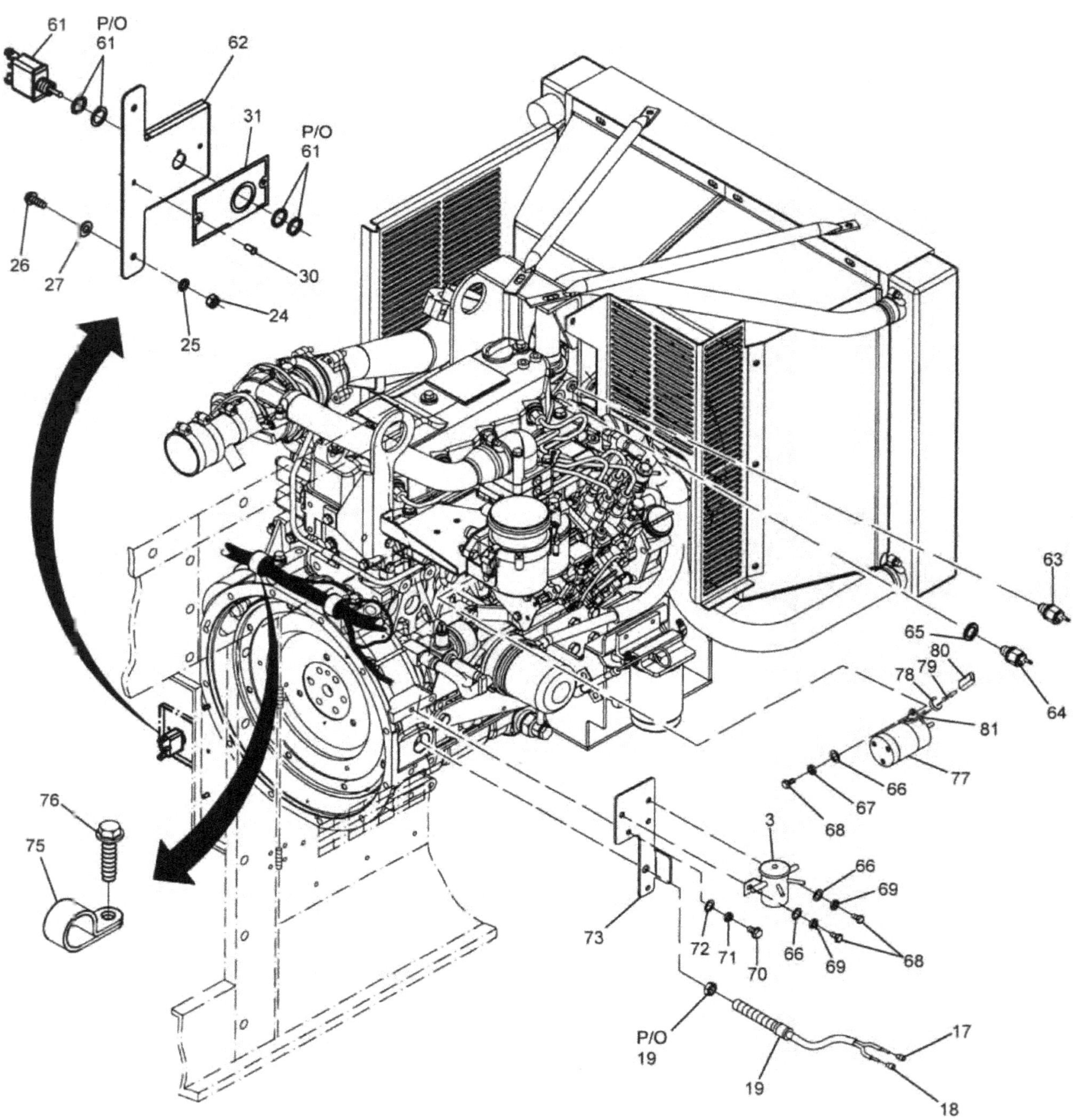

Figure 21. Engine Accessories (Sheet 5 of 5).

(1) ITEM NO	(2) SMR CODE a. ARMY	b. AIR FORCE	(3) NSN	(4) CAGEC	(5) PART NUMBER	(6) DESCRIPTION AND USABLE ON CODE (UOC)	(7) QTY

GROUP 09 ENGINE ACCESSORIES

FIG. 21 ENGINE ACCESSORIES

1 PAFZZ PAOZZ 5310-01-012-3595 30554 69-561-6 . NUT, PLAIN, ASSEMBLED UOC: EMK, YNN 3
2 PAFZZ PAOZZ 5306-01-156-7663 30554 88-20260-21 . SCREW, HEX WASHERHEAD UOC: EMK, YNN 3
3 PAFZZ PAOZZ 5945-00-855-7478 30554 88-22202 . CONTACTOR, MAGNETIC 1
4 NOT USED
5 PAFZZ PAOZZ 5305-01-380-3395 80204 B18231B10025N . SCREW, CAP, HEX HEAD UOC: EMK, YNN 2
6 PAFZZ PAOZZ 5310-01-368-3048 30554 88-22331-2 . WASHER, LOCK UOC: EMK, YNN 2
7 XDFZZ XB 30554 88-22319 . BRACKET UOC: EMK, YNN 1
8 PAFZZ PAOZZ 5930-01-107-6474 30554 88-22706 . SENDER, OIL PRESSURE 1
9 PAFZZ PAOZZ 4730-01-366-9017 30554 88-22307 . FITTING, ADAPTER 2
10 PAFZZ PAOZZ 5930-01-378-6921 30554 88-21126 . SWITCH, PRESSURE 1
11 PAFZZ PAOZZ 4820-01-367-1836 30554 88-21633 . VALVE, OIL SAMPLING 1
12 PAFZZ PAOZZ 5975-00-074-2072 96906 MS3367-1-9 . STRAP, TIEDOWN, ELECT 1
13 MDFZZ MDO 30554 88-22481 . PLATE, ID OIL SAMPL 1
14 PAFZZ PAOZZ 4730-01-236-1186 30554 88-21887 . FITTING, BRANCH TEE 1
15 PAFZZ PAOZZ 5905-01-392-8825 30554 88-22322 . SENDER, TEMPERATURE UOC: EMK, YNN 1
16 AFFFF AOO 30554 88-22550 . TRANSDUCER ASSY, MOTION (NOT SHOWN) 1
17 PAFZZ PAOZZ 5940-01-112-9746 30554 88-20275-1 . . TERMINAL, QUICK DISC 1
18 PAFZZ PAOZZ 5940-01-126-3973 30554 88-20275-2 . . TERMINAL, QUICK DISC 1
19 PAFZZ PAOZZ 6695-01-367-9723 05624 88-21061 . . TRANSDUCER, MOTION 1
20 XBFZZ XBOZZ 5310-01-500-7620 30554 88-22201 . BUSHING, BRASS UOC: EMK, YNN 1
21 PAFZZ PAOZZ 5305-01-380-3395 80204 B18231B10025N . SCREW, CAP, HEX HEAD 1
22 PAFZZ PAOZZ 5310-01-368-3048 30554 88-22331-2 . WASHER, LOCK 1
23 PAFZZ PAOZZ 5340-00-053-8994 80205 MS21333-126 . CLAMP, LOOP 1
24 PAFZZ PAOZZ 5310-00-761-6882 96906 MS51967-2 . NUT, PLAIN, HEX 2
25 PAFZZ PAOZZ 5310-00-274-8715 80205 MS35338-63 . WASHER, LOCK 2
26 PAFZZ PAOZZ 5306-01-366-7075 30554 88-20260-33 . BOLT, MACHINE 2
27 PAFZZ PAOZZ 5310-01-103-6042 96906 MS51412-4 . WASHER, FLAT 2
28 XDFZZ XB 30554 88-22320 . BRACKET, DEAD CRANK UOC: EMK, YNN 1
29 PAFZZ PAOZZ 5930-01-365-9614 30554 88-21081 . SWITCH, TOGGLE UOC: EMK, YNN 1
30 PAFZZ PAOZZ 5320-00-932-1972 97403 13214E3789-2 . RIVET, BLIND 2
31 MDFZZ MDO 30554 88-21776 . PLATE, DEAD CRANK 1
32 PAFZZ PAOZZ 5930-01-377-9113 30554 88-22317 . SWITCH, THERMOSTATIC UOC: EMK, YNN 1
33 XDFZZ XB 5305-01-423-8236 80204 B18231B10016N . SCREW, CAP, HEX HEAD 3
34 PAFZZ PAOZZ 5310-01-368-3048 30554 88-22331-2 . WASHER, LOCK 2
35 PAFZZ PAOZZ 5340-00-057-3026 80205 MS21333-117 . CLAMP, LOOP 1
36 PAFZZ PAOZZ 4730-00-287-3281 81343 2-130109E . PLUG, PIPE 1

(1)	(2) SMR CODE		(3)	(4)	(5)	(6)	(7)
	a.	b.					
ITEM NO	ARMY	AIR FORCE	NSN	CAGEC	PART NUMBER	DESCRIPTION AND USABLE ON CODE (UOC)	QTY

37 PAFZZ PAOZZ 80204 B1831BH060014 . SCREW, CAP, SOCKET HEAD 4
38 PAFZZ PAOZZ 5310-01-195-6611 30554 88-22331-3 . WASHER, LOCK 10
39 PBFZZ PBOZZ 5340-01-367-2287 30554 88-22623 . CLEVIS, ROD END 1
40 PBFZZ PBOZZ 5340-01-476-5376 30554 88-22784 . CLIP, SPRING TENSION 1
41 PAFZZ PAOZZ 5310-01-220-6513 80204B18241A06088-22545 . NUT, PLAIN, HEX .... 1 42 AFFFF AOO 1440-01-560-1702 30554 . ACTUATOR ASSY (NOT SHOWN)
UOC: EMK, YNN 1
43 PAFZZ PAOZZ 5975-00-074-2072 96906 MS3367-1-9 . . STRAP, TIEDOWN, ELECT
UOC: EMK, YNN 1
44 PAFZZ PAOZZ 5999-01-092-2655 30554 88-20476 . . CONTACT, ELECTRICAL
UOC: EMK, YNN 2
45 PAFZA PAOZA 5935-00-482-7721 30554 88-20471 . . CONNECTOR, PLUG, ELEC
UOC: EMK, YNN 1
46 PAFZZ PAOZZ 2910-01-371-2689 30554 88-22428 . . ACTUATOR, ELECTRO-ME
UOC: EMK, YNN 1
47 PAFZZ PAOZZ 5330-01-366-6589 30554 88-22618 . GASKET 1
48 PAFZZ PAOZZ 80204 B18231B6020N . SCREW, CAP, SOCKET HE 6
49 XDFZZ XB 30554 88-22617 . PLATE, INTERFACE 1
50 PAFZZ PAOZZ 5330-01-392-8826 30554 88-22416 . GASKET 1
51 PAFZZ PAOZZ 5310-01-366-3540 80204 B18241B050 . NUT, PLAIN HEX 1
52 PAFZZ PAOZZ 5310-01-368-3049 30554 88-22331-4 . WASHER, LOCK 1
53 PAFZZ PAOZZ 80204 B18231B5016N . SCREW, CAP, HEX HEAD 1
54 PAFZZ PAOZZ 5340-01-367-2328 30554 88-22619 . BRACKET, ANGLE 1
55 PAFZZ PAOZZ 5310-00-208-9255 30554 95-8125-4 . NUT, PLAIN, HEX 2
56 XDFZZ XB 30554 88-22621 . ADAPTER, RACK 1
57 PBFZZ PBOZZ 5315-01-366-6685 30554 88-22620 . PIN, STRAIGHT, THREAD 1
58 PBFZZ PBOZZ 5360-01-368-4663 30554 88-22624 . SPRING, HELICAL, COMP .... 1 59 PAFZZ PAOZZ 5310-01-257-7590 96906 MS51412-7 . WASHER, FLAT
UOC: EMK 3
60 PAFZZ PAOZZ 5310-00-988-2652 80205 MS35650-103 . NUT, PLAIN, HEX .... 2 61 PAFZZ PAFZZ 5930-01-365-9614 30554 88-21081 . SWITCH, TOGGLE, DEAD CRANK
UOC: 86Q, 86R............................. 1
62 XBFZZ XBFZZ 30554 97-24045 . BRACKET, DEAD CRANK SWITCH
UOC: 86Q, 86R............................. 1
63 PAFZZ PAFZZ 30554 97-24061 . SWITCH, THERMOSTATIC
UOC: 86Q, 86R............................. 1
64 PAFZZ PAFZZ 30554 88-22322 . SENDER, TEMPERATURE
UOC: 86Q, 86R............................. 2
65 PAFZZ PAFZZ 30554 97-24118 . WASHER, COPPER
UOC: 86Q, 86R............................. 1
66 PAFZZ PAFZZ 5310-01-531-4939 30554 88-20033-22A . WASHER, FLAT
UOC: 86Q, 86R............................. 4
67 PAFZZ PAFZZ 5310-01-195-6611 30554 88-22331-3 . WASHER, LOCK
UOC: 86Q, 86R............................. 3
68 PAFZZ PAFZZ 5305-01-392-8847 81343 B1831BH060016 . SCREW
UOC: 86Q, 86R............................. 4
69 PAFZZ PAFZZ 5310-01-466-7321 30554 88-22792-2 . WASHER, LOCK
UOC: 86Q, 86R............................. 1
70 PAFZZ PAFZZ 5306-01-179-1438 80204 B18231A10025N . BOLT, MACHINE
UOC: 86Q, 86R............................. 2
71 PAFZZ PAFZZ 5310-01-099-9532 30554 88-22331-1 . WASHER, LOCK
UOC: 86Q, 86R............................. 2
72 PAFZZ PAFZZ 5310-01-531-4937 30554 88-20033-24A . WASHER, FLAT
UOC: 86Q, 86R............................. 2

(1)	(2) SMR CODE a.	b.	(3)	(4)	(5)	(6)	(7)
ITEM NO	ARMY	AIR FORCE	NSN	CAGEC	PART NUMBER	DESCRIPTION AND USABLE ON CODE (UOC)	QTY
73	XBFZZ	XBFZZ		30554	97-24017	. BRACKET ASSEMBLY UOC: 86Q, 86R	1
74	NOT USED						
75	PAFZZ	PAFZZ		30554	88-20554-1	. CLAMP, LOOP UOC: 86Q, 86R	2
76	PAFZZ	PAFZZ	5306-01-174-8738	80204	B18231A10020N	. SCREW, CAP, SOCKET HEAD UOC: 86Q, 86R	2
77	AFFFF	AOO		30554	97-24114	. ACTUATOR ASSY (NOT SHOWN) UOC: 86Q, 86R	1
78	PAFZZ	PAOZZ	5975-00-074-2072	96906	MS3367-1-9	. . STRAP, TIEDOWN, ELECT UOC: 86Q, 86R	1
79	PAFZZ	PAFZZ	5999-01-092-2655	30554	88-20476	. . CONTACT, ELECTRICAL UOC: 86Q, 86R	2
80	PAFZA	PAOZA	5935-00-482-7721	27264	88-20471	. . CONNECTOR, PLUG, ELEC UOC: 86Q, 86R	1
81	PAFZZ	PAFZZ		30554	97-24018	. . ACTUATOR, GOVERNOR UOC: 86Q, 86R	1

END OF FIGURE

FIELD AND SUSTAINMENT MAINTENANCE

15 kW 50/60 AND 400 Hz SKID MOUNTED TACTICAL QUIET GENERATOR SETS

GROUP 10 LUBRICATION SYSTEM: OIL DRAIN LINE

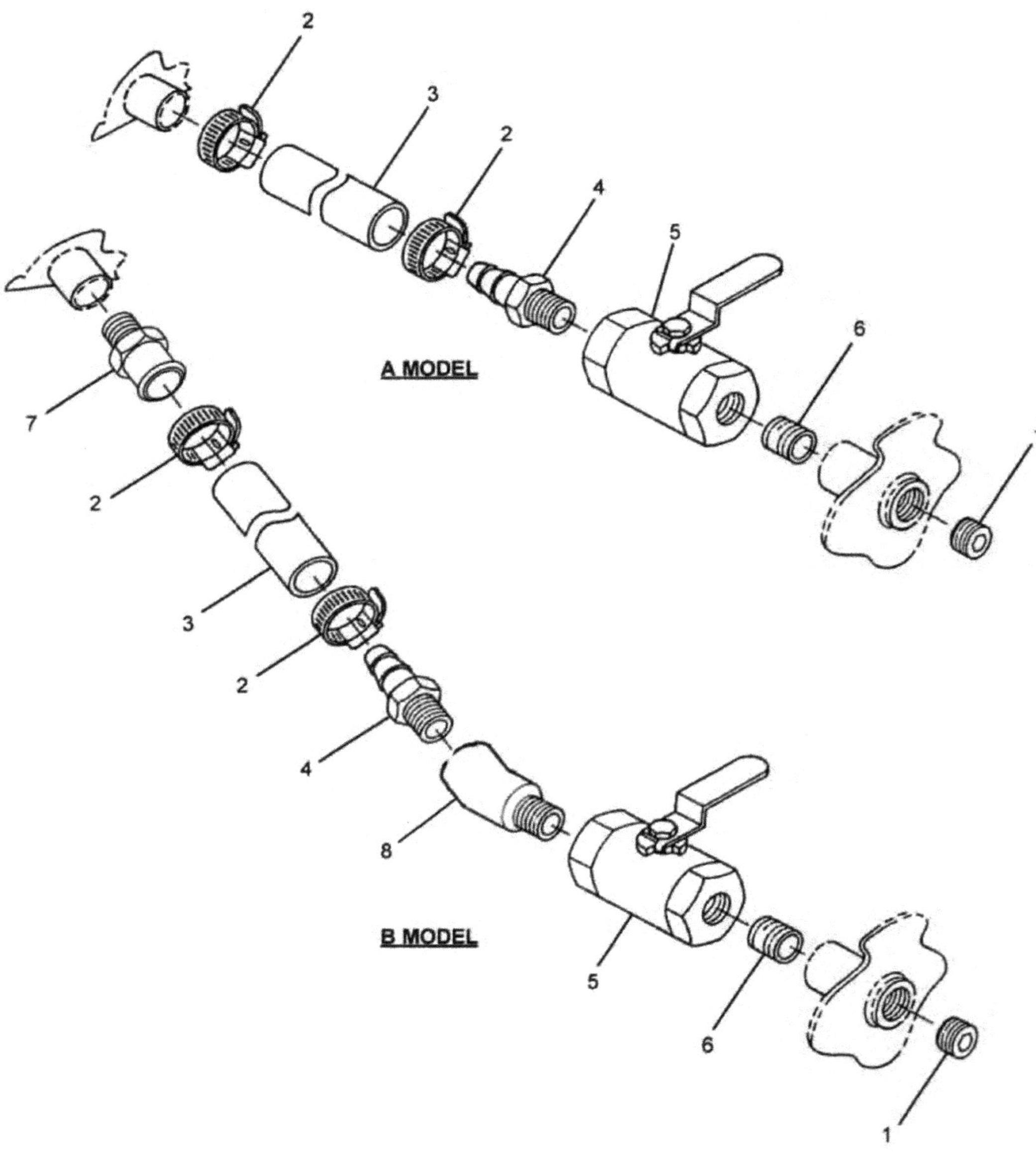

Figure 22. Oil Drain Line.

(1)	(2)		(3)	(4)	(5)	(6)	(7)
	SMR CODE						
	a.	b.					
ITEM NO	ARMY	AIR FORCE	NSN	CAGEC	PART NUMBER	DESCRIPTION AND USABLE ON CODE (UOC)	QTY

GROUP 10 LUBRICATION SYSTEM

FIG. 22 OIL DRAIN LINE

1 PAFZZ PAOZZ 30554 12 130109G . PLUG, PIPE 1

2 PAFZZ PAOZZ 4730-00-908-3194 30554 88-20561-2 . CLAMP, HOSE .... 2 3 MFFZZ MOO 30554 88-21604-110 . HOSE, OIL DRAIN MAKE FROM P/N

100R1A-16 (81349), 24 INCHES REQUIRD
UOC: EMK 1

3 MFFZZ MOO 30554 97-24604-110 . . HOSE, OIL DRAIN MAKE FROM P/N

100R1AT-16 (81349), 24 INCHES REQUIRED
UOC: 86Q.................................. 1

3 MFFZZ MOO 30554 88-21605-110 . HOSE, OIL DRAIN MAKE FROM P/N

100R1A-16 (81349), 24 INCHES REQUIRED
UOC: YNN 1

3 MFFZZ MOO 30554 97-24605-110 . . HOSE, OIL DRAIN MAKE FROM P/N

100R1AT-16 (81349), 24 INCHES REQUIRED
UOC: 86R 1

4 PAFZZ PAOZZ 30554 88-21755-3 . ADAPTER, STRAIGHT, PI 1

5 PAFZZ PAOZZ 30554 88-21891 . VALVE, BALL 1

6 PAFZZ PAOZZ 4730-01-470-1982 30554 72-5304 . NIPPLE, PIPE .... 1 7 PAFZZ PAFZZ 30554 97-24011 . FITTING, OIL DRAIN

UOC: 86Q, 86R............................. 1

8 PAFZZ PAFZZ 4730-00-900-8663 81343 12-12 140339C . ELBOW 45' ST 3/4 NPTX 3/4 FEMALE NPTSAEJ516
UOC: 86Q, 86R............................. 1

END OF FIGURE

FIELD AND SUSTAINMENT MAINTENANCE

15 kW 50/60 AND 400 Hz SKID MOUNTED TACTICAL QUIET GENERATOR SETS

GROUP 10 LUBRICATION SYSTEM: LUBRICATION ACCESSORIES

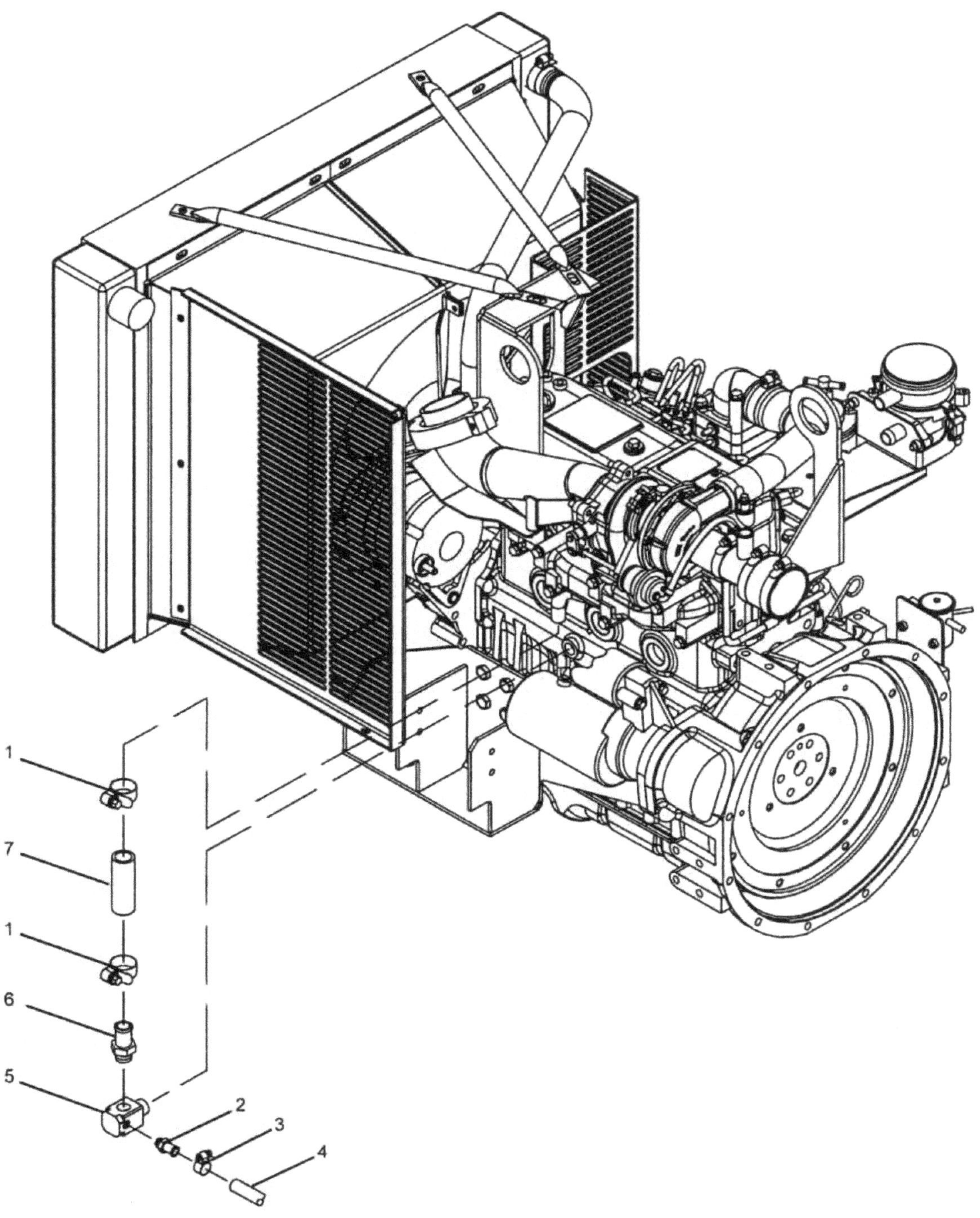

Figure 23. Lubrication Accessories (Sheet 1 of 2).

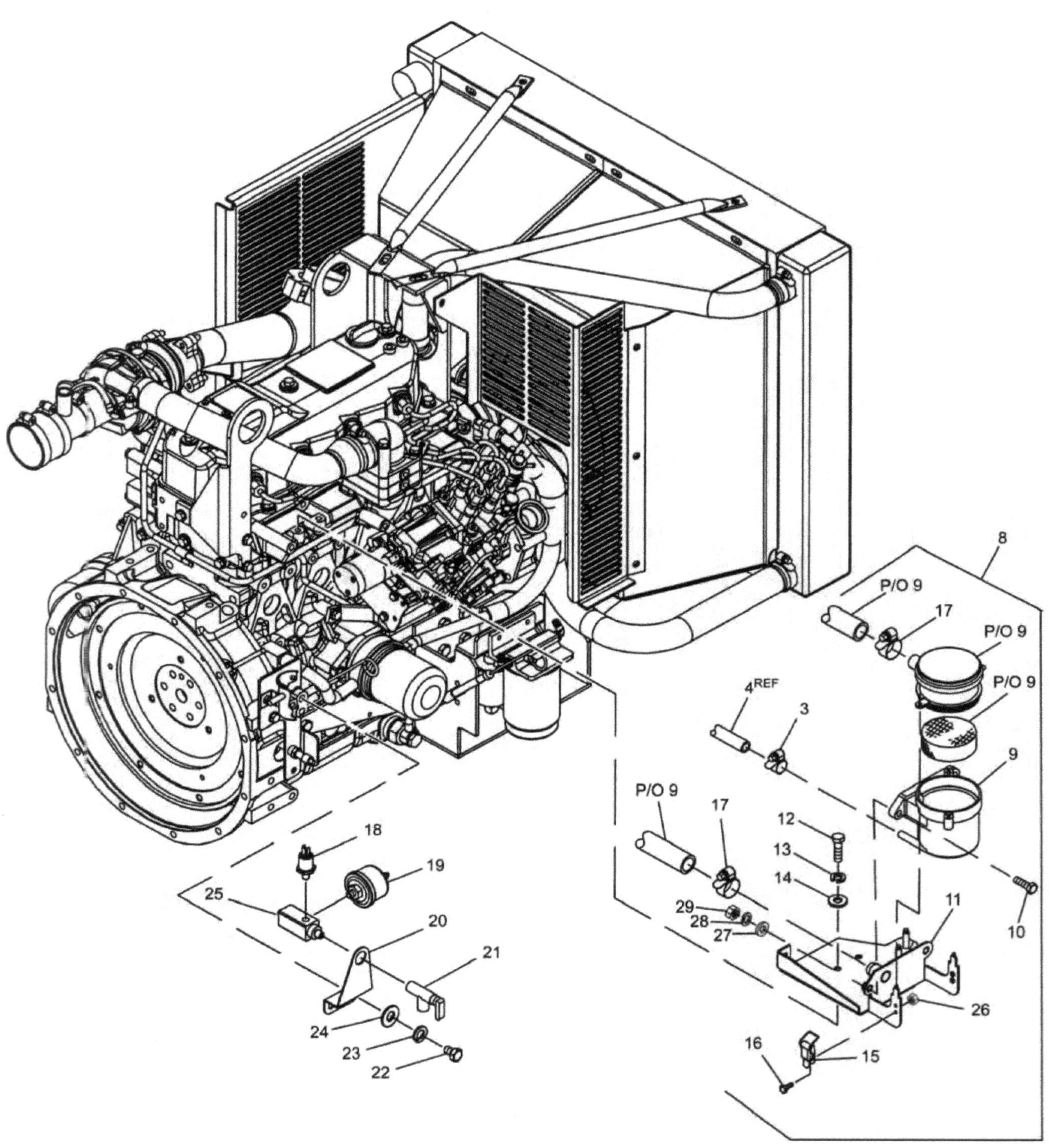

Figure 23. Lubrication Accessories (Sheet 2 of 2).

(1) ITEM NO	(2) SMR CODE a. ARMY	b. AIR FORCE	(3) NSN	(4) CAGEC	(5) PART NUMBER	(6) DESCRIPTION AND USABLE ON CODE (UOC)	(7) QTY
						GROUP 10 LUBRICATION SYSTEM	
						FIG. 23 LUBRICATION ACCESSORIES	
1	PAFZZ	PAFZZ	4730-01-470-2409	30554	88-20561-2	. CLAMP, HOSE UOC: 86Q, 86R	2
2	PAFZZ	PAFZZ	4730-01-417-5855	81343	8-4 430160C	. ADAPTER, STRAIGHT UOC: 86Q, 86R	2
3	PAFZZ	PAFZZ	4730-01-470-1626	30554	88-20561-1	. CLAMP, HOSE UOC: 86Q, 86R	2
4	MFFZZ	MOF	4720-00-670-6037	30554	97-24604-282	. HOSE, NONMETALLIC, MAKE FROM P/N 88-20579-6 (30554), 27 INCHES REQUIRED UOC: 86Q	1
4	MFFZZ	MOF		30554	97-24605-282	. HOSE, NONMETALLIC, MAKE FROM P/N 88-20579-6 (30554), 27 INCHES REQUIRED UOC: 86R	1
5	PAFZZ	PAFZZ		30554	97-24031	. FITTING, TEE, OIL DRAIN UOC: 86Q, 86R	1
6	PAFZZ	PAFZZ		30554	97-24113	. CONNECTOR UOC: 86Q, 86R	1
7	MFFZZ	MOF		30554	97-24604-112	. HOSE, NONMETALLIC, MAKE FROM P/N 88-20579-7 (30554), 2.75 INCHES REQUIRED UOC: 86Q	1
7	MFFZZ	MOF		30554	97-24605-112	. HOSE, NONMETALLIC, MAKE FROM P/N 88-20579-7 (30554), 2.75 INCHES REQUIRED UOC: 86Q	1
8	XBFFF	XBFFF		30554	97-24002	. FILTER ASSEMBLY, CRANKCASE VENTILATION (NOT SHOWN) UOC: 86Q, 86R	1
9	PAFZZ	PAFZZ		30554	97-24025	. . FILTER, CRANKCASE VENTILA- TION UOC: 86Q, 86R	1
10	PAFZZ	PAFZZ	5306-00-226-4827	80204	B1821BH031C100N	. . SCREW, HEX HEAD UOC: 86Q, 86R	2
11	XBFFF	XBFFF		30554	97-24024	. . BRACKET ASSEMBLY, FILTER UOC: 86Q, 86R	1
12	PAFZZ	PAFZZ	5306-01-179-1438	80204	B18231A1025N	. BOLT, MACHINE UOC: 86Q, 86R	2
13	PAFZZ	PAFZZ	5310-01-099-9532	30554	88-22331-1	. WASHER, LOCK UOC: 86Q, 86R	2
14	PAFZZ	PAFZZ	5310-01-531-4937	30554	88-20033-24A	. WASHER, FLAT UOC: 86Q, 86R	2
15	PAFZZ	PAFZZ		30554	97-24026	. . LATCH, TOGGLE UOC: 86Q, 86R	2
16	PAFZZ	PAFZZ	5305-00-036-6972	30554	69-662-20	. . SCREW, HEX HEAD UOC: 86Q, 86R	4
17	PAFZZ	PAFZZ	4730-01-470-1567	30554	88-20561-3	. CLAMP, HOSE UOC: 86Q, 86R	4
18	PAFZZ	PAFZZ	5930-01-378-6921	30554	88-21126	. SWITCH, PRESSURE UOC: 86Q, 86R	1

(1) ITEM NO	(2) SMR CODE a. ARMY	b. AIR FORCE	(3) NSN	(4) CAGEC	(5) PART NUMBER	(6) DESCRIPTION AND USABLE ON CODE (UOC)	(7) QTY
19	PAFZZ	PAFZZ	5930-01-107-6474	30554	88-22706	. SENSOR, OIL PRESSURE UOC: 86Q, 86R	1
20	XBFFF	XBFFF		30554	97-24016	. BRACKET, MANIFOLD UOC: 86Q, 86R	1
21	PAFZZ	PAFZZ	4820-01-367-1836	30554	88-21633	. VALVE, OIL SAMPLING UOC: 86Q, 86R	1
22	PAFZZ	PAFZZ	5306-01-174-8738	30554	B18231A100020N	. BOLT, MACHINE UOC: 86Q, 86R	2
23	PAFZZ	PAFZZ	5310-01-368-3048	30554	88-22331-2	. WASHER, LOCK UOC: 86Q, 86R	2
24	PAFZZ	PAFZZ	5310-01-532-0321	30554	88-20033-31A	. WASHER, FLAT UOC: 86Q, 86R	2
25	PAFZZ	PAFZZ		30554	97-24015	. MANIFOLD, OIL PRESSURE UOC: 86Q, 86R	1
26	PAFZZ	PAFZZ	5310-00-063-7360	30554	69-561-2	. . NUT, CAGE UOC: 86Q, 86R	4
27	PAFZZ	PAFZZ	5310-01-234-9415	30554	88-20564-3	. . WASHER, FLAT UOC: 86Q, 86R	2
28	PAFZZ	PAFZZ	5310-01-477-9621	30554	88-20556-6	. . WASHER, LOCK UOC: 86Q, 86R	2
29	PAFZZ	PAFZZ	5310-01-470-1286	30554	88-22790-2	. . NUT UOC: 86Q, 86R	2

END OF FIGURE

FIELD AND SUSTAINMENT MAINTENANCE

15 kW 50/60 AND 400 Hz SKID MOUNTED TACTICAL QUIET GENERATOR SETS

GROUP 11 GENERATOR INSTALLATION: GENERATOR ASSEMBLY

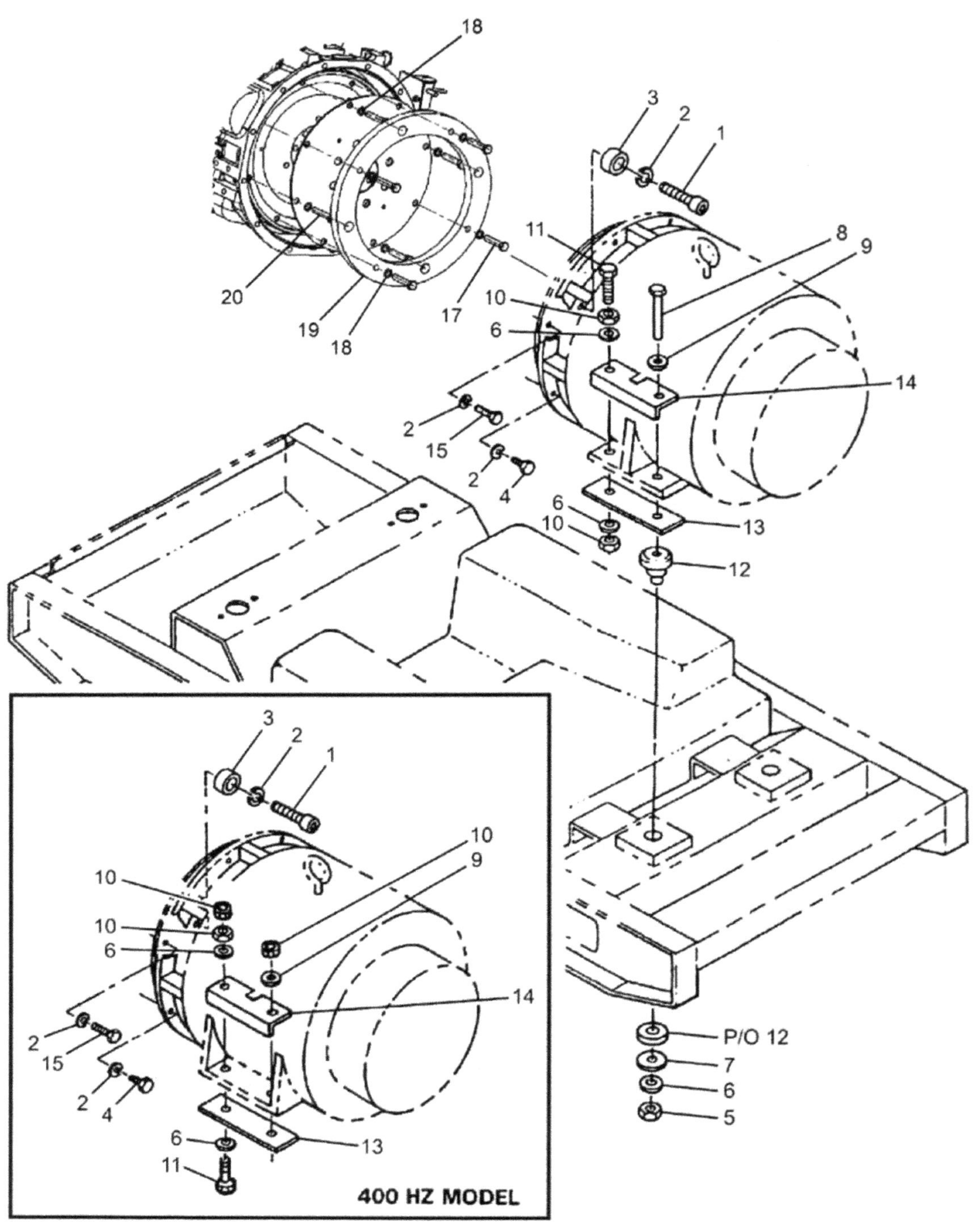

Figure 24. Generator Assembly.

(1)	(2)		(3)	(4)	(5)	(6)	(7)
	SMR CODE						
	a.	b.					
ITEM NO	ARMY	AIR FORCE	NSN	CAGEC	PART NUMBER	DESCRIPTION AND USABLE ON CODE (UOC)	QTY

GROUP 11 GENERATOR INSTALLATION

FIG. 24 GENERATOR ASSEMBLY

1 PAFZZ PAOZZ 5305-00-983-7449 96906 MS16998-75 . SCREW, CAP, SOCKET HEAD 2
2 PAFZZ PAOZZ 5310-00-011-5093 80205 MS35338-65 . WASHER, LOCK 22
3 PAFZZ PAOZZ 5365-01-384-9169 30554 88-22339 . SPACER, SLEEVE 2
4 PAFZZ PAOZZ 5305-00-068-0511 80204 B1821BH038C125N . SCREW, CAP, HEX HEAD 12
5 PAFZZ PAOZZ01-215-7311310-01-215-731196906 MS51943-13 . NUT, SELF-LOCKING, HEX .... 2 6 PAFZZ PAOZZ 5310-01-242-2679 96906MS51412-11 .WASHER,
FLAT 6
7 PAFZZ PAOZZ 5310-01-365-8139 30554 88-20247-3 . WASHER, FLAT 2
8 PAFZZ PAOZZ 5305-00-724-7265 80204 B1821BH063C475N . SCREW, CAP, HEX HEAD 2
9 PBFZZ PBOZZ 5310-01-365-4386 92830 B1875-127 . WASHER, BEVEL 2
10 PAFZZ PAOZZ 5310-00-764-6609 96906 MS51971-7 . NUT, PLAIN, HEX 4
11 PAFZZ PAOZZ 5305-01-366-3501 30554 88-22530 . SCREW, CAP, HEX HEAD 2
12 PBFZZ PBOZZ 5340-01-366-3360 81860 RB-X85 . MOUNT, RESILIENT 2
13 XDFZZ XB 30554 88-22524 . PLATE, GENERATOR MO 2
14 XDFZZ XB 30554 88-22525 . ANGLE, GENERATOR 2
15 PAFZZ PAOZZ 5305-00-068-0510 80204B1821BH038 C100N . SCREW, CAP, HEX HEAD .... 8 16 NOT USED
17 PAFZZ PAOZZ 30554 97-24116-5 . SCREW, CAP, HEX HEAD
UOC: 86Q, 86R............................. 4
18 PAFZZ PAOZZ 5310-01-368-3048 30554 88-22331-2 . WASHER, LOCK
UOC: 86Q, 86R............................. 8
19 PAFZZ PAOZZ 30554 97-24122 . RING, INERTIA
UOC: 86Q, 86R............................. 1
20 PAFZZ PAFZZ 30554 97-24116-4 . SCREW, CAP, HEX HEAD
UOC: 86Q, 86R............................. 4

END OF FIGURE

FIELD AND SUSTAINMENT MAINTENANCE

15 kW 50/60 AND 400 Hz SKID MOUNTED TACTICAL QUIET GENERATOR SETS

GROUP 11 GENERATOR ASSEMBLY: GENERATOR ASSEMBLY, 50/60 HZ

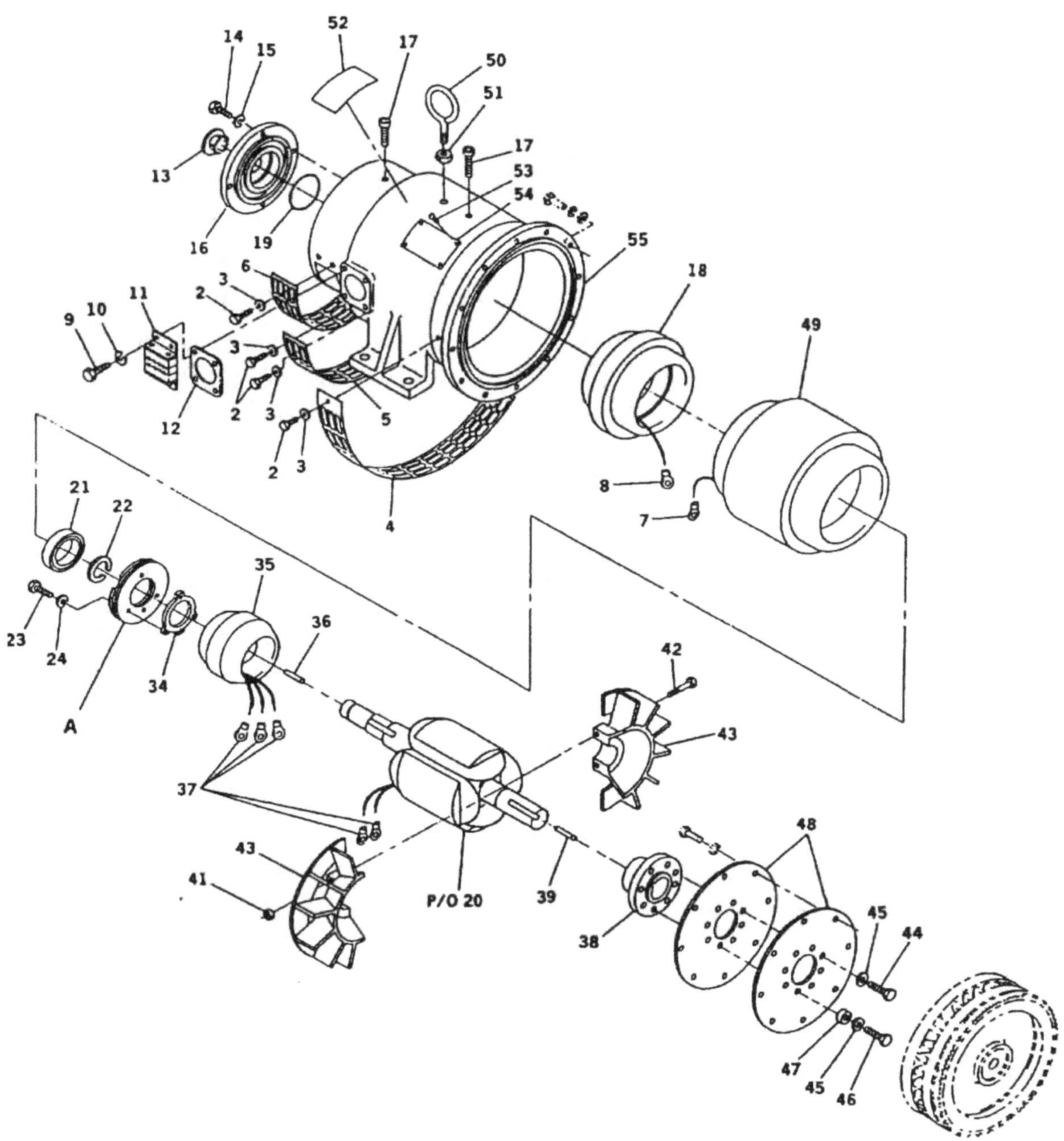

Figure 25. Generator Assembly, 50/60 Hz (Sheet 1 of 2).

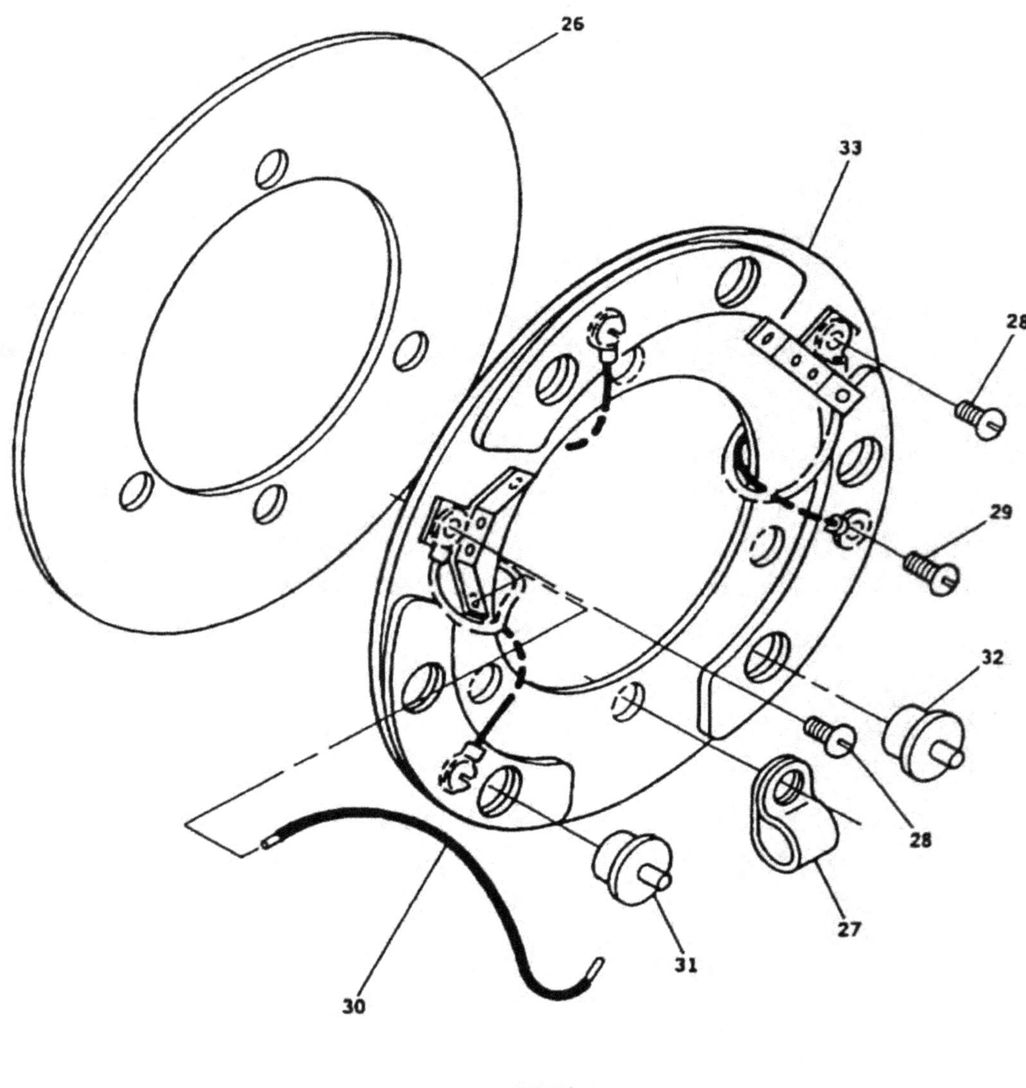

Figure 25. Generator Assembly, 50/60 Hz (Sheet 2 of 2).

(1)	(2) SMR CODE a.	b.	(3)	(4)	(5)	(6)	(7)
ITEM NO	ARMY	AIR FORCE	NSN	CAGEC	PART NUMBER	DESCRIPTION AND USABLE ON CODE (UOC)	QTY
						GROUP 11 GENERATOR ASSEMBLY	
						FIG. 25 GENERATOR ASSEMBLY, 50/60 HZ	
1	PBFFH	PBFFH	6115-01-383-3124	30554	88-21005	. GENERATOR, ALTERNATING CURRENT, 60 HZ (NOT SHOWN) UOC: EMK, 86Q	1
2	PAFZZ	PAOZZ	5305-00-068-0500	96906	MS90725-3	. . SCREW, CAP, HEX HEAD UOC: EMK, 86Q	10
3	XDFZZ	XB	5310-01-215-7311	96906	MS51943-13	. . WASHER, FLAT UOC: EMK, 86Q	10
4	XDFZZ	XB		36156	720346-0C	. . COVER, BAND UOC: EMK, 86Q	1
5	XDFZZ	XB		36156	720345-03	. . SCREEN, INTAKE UOC: EMK, 86Q	1
6	XDFZZ	XB		36156	720344-03	. . SCREEN, INTAKE UOC: EMK, 86Q	1
7	PAFZZ	PAOZZ	5940-00-113-9820	96906	MS20659-128	. . TERMINAL, LUG USED ON STATOR LEADS UOC: EMK, 86Q	10
8	PAFZZ	PAOZZ	5940-00-113-0954	96906	MS20659-165	. . TERMINAL, LUG USED ON EXCITER LEADS UOC: EMK, 86Q	2
9	PAFZZ	PAOZZ	5305-00-068-0502	80205	MS90725-6	. . SCREW, CAP, HEX HEAD UOC: EMK, 86Q	4
10	PAFZZ	PAOZZ	5310-00-274-8715	80205	MS35338-63	. . WASHER, LOCK UOC: EMK, 86Q	4
11	XDFZZ	XB		36156	752818-0A	. . LEAD CLAMP ASSEMBLY UOC: EMK, 86Q	1
12	XDFZZ	XB		36156	834822-01	. . GASKET, LEAD CLAMP UOC: EMK, 86Q	1
13	PAFZZ	PAOZZ	5340-01-382-2589	36156	846833-01	. . PLUG, PROTECTIVE DUS UOC: EMK, 86Q	1
14	PAFZZ	PAOZZ	5306-00-225-8499	80205	MS90725-34	. . SCREW, CAP, HEX HEAD UOC: EMK, 86Q	4
15	PAFZZ	PAOZZ	5310-00-011-6120	96906	MS35338-64	. . WASHER, LOCK UOC: EMK, 86Q	4
16	XDFZZ	XB		36156	703320-02	. . ENDBELL UOC: EMK, 86Q	1
17	PAFZA	PAFZA	5305-00-292-4562	88044	AN565C416-12	. . SETSCREW UOC: EMK, 86Q	2
18	XDFZZ	XB		36156	789295-0A	. . STATOR CORE, EXCITE UOC: EMK, 86Q	1
19	PAFZZ	PAOZZ	5330-01-374-4468	36156	865876-01	. . O-RING UOC: EMK, 86Q	1
20	XDFFF	XB		36156	777070-0A	. . ROTOR ASSY 60 HZ (NOT SHOWN) UOC: EMK, 86Q	1
21	PAFZZ	PAOZZ	3110-01-304-8142	52676	6308-2RSJEM	. . . BEARING, BALL, ANNULA UOC: EMK, 86Q	1
22	XDFZZ	XB	5365-00-803-7304	96906	MS16624-1156	...RING, RETAINING UOC: EMK, 86Q	1

(1) ITEM NO	(2) SMR CODE a. ARMY	b. AIR FORCE	(3) NSN	(4) CAGEC	(5) PART NUMBER	(6) DESCRIPTION AND USABLE ON CODE (UOC)	(7) QTY
23	PAFZZ	PAOZZ	5305-01-324-4896	36156	801016-06	...SCREW, CAP, HEXHEAD UOC: EMK, 86Q	3
24	PAFZZ	PAOZZ	5310-00-576-5752	80205	MS35333-39	...WASHER,LOCK UOC: EMK, 86Q	3
25	PAFZZ	PAOZZ	5961-01-248-1712	36156	777079-0A	. . . SEMICONDUCTOR DEVICE (NOT SHOWN) UOC: EMK, 86Q	1
26	XDFZZ	XB		36156	718810	...PLATE,RECTIFIER INSULATING UOC: EMK, 86Q	1
27	PAFZZ	PAOZZ	5340-01-302-2960	96906	MS25281-4	...CLAMP, LOOP UOC: EMK, 86Q	1
28	PAFZZ	PAOZZ	5305-00-984-6189	80205	MS35206-241	...SCREW, MACHINE UOC: EMK, 86Q	3
29	PAFZZ	PAOZZ	5305-00-993-1849	80205	MS35207-259	...SCREW, MACHINE UOC: EMK, 86Q	2
30	MFFZZ	MOO		30554	88-21795-8	. . . WIRE, ELECTRICAL MAKE FROM P/N 88-20444-2 (30554), 14 GA WIRE AS REQUIRED UOC: EMK, 86Q	2
31	PAFZZ	PAOZZ	5961-00-478-7687	80131	1N3663	. . . SEMICONDUCTOR DEVICE DIODE 30 AMP, 400 VOLT UOC: EMK, 86Q	3
32	PAFZZ	PAOZZ	5961-01-013-0682	80121	1N3663R	. . . SEMICONDUCTOR DEVICE DIODE REVERSE 30 AMP, 400 VOLTS UOC: EMK, 86Q	3
33	XAFZZ	XA		36156	778715-0A	...PLATE,RECTIFIER MO UOC: EMK, 86Q	1
34	XDFZZ	XB		36156	707806-0A	...RECTIFIER, HUB UOC: EMK, 86Q	1
35	XDFZZ	XB		36156	791149-0A	...ROTOR ASSY, EXCITER UOC: EMK, 86Q	1
36	XDFZZ	XB	5315-01-236-0612	96906	MS20066-352	. . . KEY, MACHINE UOC: EMK, 86Q	1
37	PAFZZ	PAOZZ	5940-00-114-1300	96906	MS20659-105	...TERMINAL, LUG UOC: EMK, 86Q	5
38	XDFZZ	XB		36156	707329-01	. . . HUB, DRIVE UOC: EMK, 86Q	1
39	XDFZZ	XB	5315-00-227-6415	96906	MS20066-354	. . . KEY, MACHINE UOC: EMK, 86Q	1
40	PBFZZ	PBOZZ	2930-01-329-4857	36156	716327-0A	...IMPELLER, FAN, AXIAL (NOT SHOWN) UOC: EMK, 86Q	1
41	PAFZZ	PAOZZ	5310-00-984-3806	96906	MS51922-9	....NUT,SELF-LOCKING, HEX UOC: EMK, 86Q	2
42	PAFZZ	PAOZZ	5305-00-225-8507	80205	MS90725-43	....SCREW, CAP, HEXHEAD UOC: EMK, 86Q	2
43	XAFZZ	XA		30554	88-21783	....FAN,HALF UOC: EMK, 86Q	2
44	PAFZZ	PAOZZ	5305-00-269-2801	96906	MS90726-58	. . SCREW, CAP, HEX HEAD UOC: EMK, 86Q	6
45	PAFZZ	PAOZZ	5310-00-087-7493	96906	MS27183-13	. . WASHER, FLAT UOC: EMK, 86Q	8
46	PAFZZ	PAOZZ	5305-00-269-3239	80204	B1821BH038F138N	. . SCREW, CAP, HEX HEAD UOC: EMK, 86Q	2
47	XDFZZ	XB	5365-01-384-9169	30554	88-22339	. . SPACER, SLEEVE UOC: EMK, 86Q	2

(1) ITEM NO	(2) SMR CODE a. ARMY	b. AIR FORCE	(3) NSN	(4) CAGEC	(5) PART NUMBER	(6) DESCRIPTION AND USABLE ON CODE (UOC)	(7) QTY
48	XDFZZ	XB		36156	702806-01	. . DISC, COUPLING UOC: EMK, 86Q	2
49	XDFZZ	XB		36156	783621-0A	. . STATOR ASSY UOC: EMK, 86Q	1
50	PAFZZ	PAOZZ	5306-00-050-0347	96906	MS51937-5	. . BOLT, EYE UOC: EMK, 86Q	1
51	PAFZZ	PAOZZ	5310-00-768-0318	96906	MS51967-14	. . NUT, PLAIN, HEX UOC: EMK, 86Q	1
52	MDFZZ	MDO		36156	A-9696	. . LABEL, WARNING UOC: EMK, 86Q	1
53	PAFZZ	PAOZZ	5305-00-253-5615	80205	MS21318-21	. . SCREW, DRIVE UOC: EMK, 86Q	4
54	MDFZZ	MDO		30554	88-20064-5	. . PLATE, IDENTIFICATION UOC: EMK, 86Q	1
55	XDFZZ	XB		36156	706336-03	. . FRAME, STATOR HOUSING UOC: EMK, 86Q	1

END OF FIGURE

FIELD AND SUSTAINMENT MAINTENANCE

15 kW 50/60 AND 400 Hz SKID MOUNTED TACTICAL QUIET GENERATOR SETS

GROUP 11 GENERATOR ASSEMBLY: GENERATOR ASSEMBLY, 400 HZ

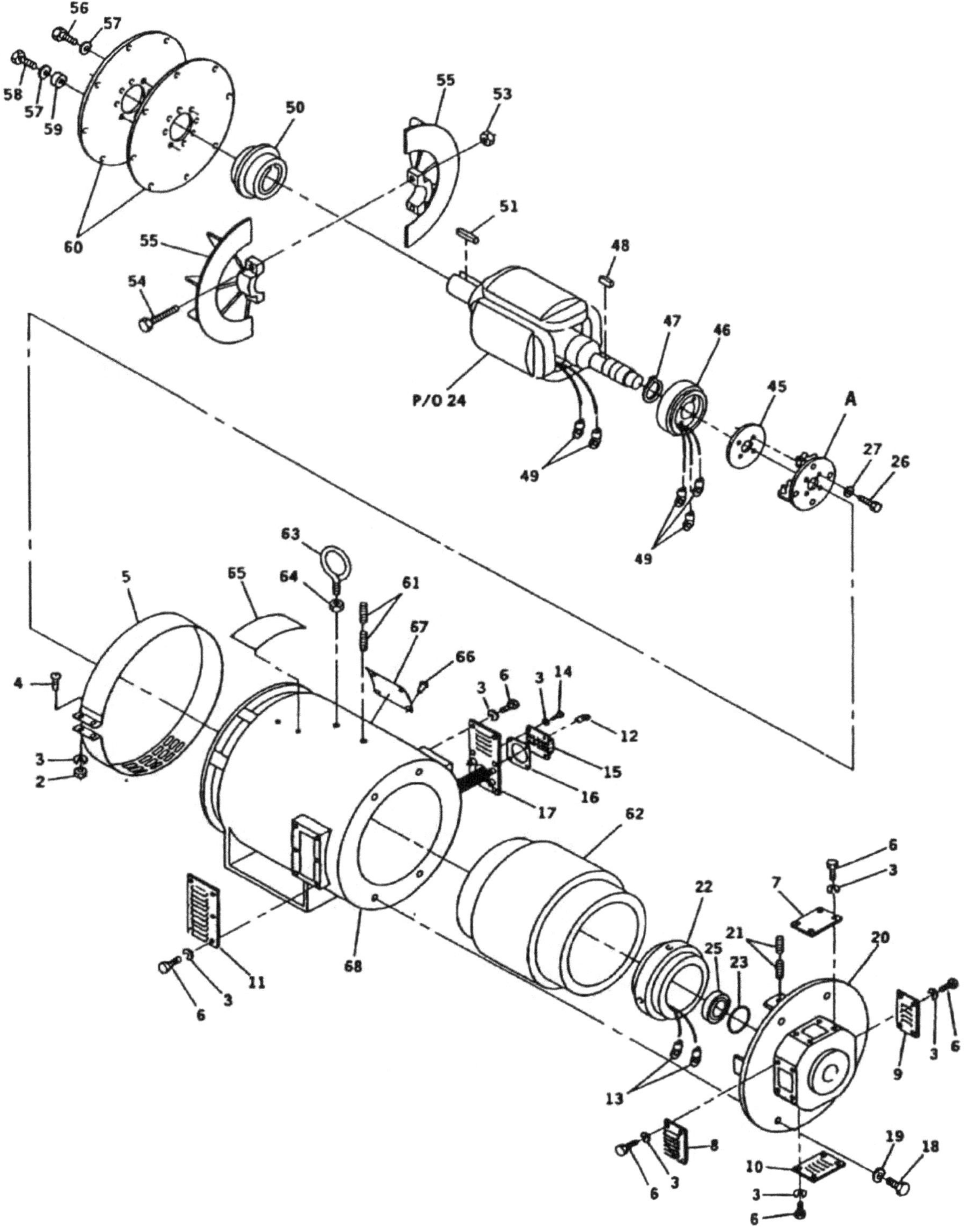

Figure 26. Generator Assembly, 400 Hz (Sheet 1 of 2).

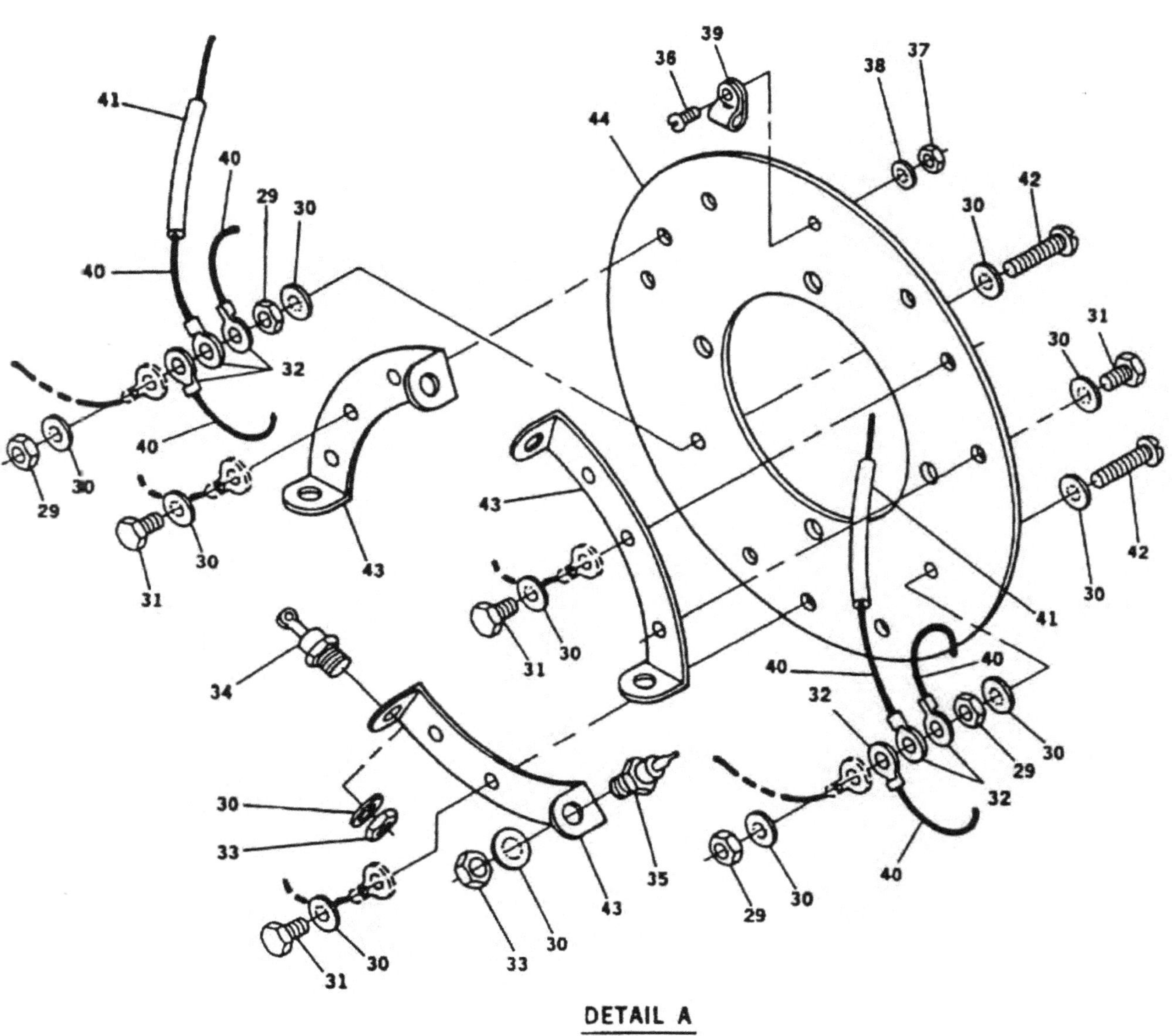

Figure 26. Generator Assembly, 400 Hz (Sheet 2 of 2).

(1) ITEM NO	(2) SMR CODE a. ARMY	b. AIR FORCE	(3) NSN	(4) CAGEC	(5) PART NUMBER	(6) DESCRIPTION AND USABLE ON CODE (UOC)	(7) QTY
						GROUP 11 GENERATOR ASSEMBLY	
						FIG. 26 GENERATOR ASSEMBLY, 400 HZ	
1	PBFFH	PBFFH	6115-01-434-1432	30554	88-21006	. GENERATOR, ALTERNATING CURRENT, 400 HZ (NOT SHOWN) UOC: YNN, 86R	1
2	PAFZZ	PAOZZ00-761-68825310-00-761-688296906			MS51967-2	. . NUT, PLAIN, HEX UOC: YNN, 86R	2
3	PAFZZ	PAOZZ	5310-00-274-8715	80205	MS35338-63	. . WASHER, LOCK UOC: YNN, 86R	34
4	PAFZZ	PAOZZ	5305-00-988-1728	80205	MS35206-287	. . SCREW, MACHINE UOC: YNN, 86R	2
5	XDFZZ	XB		36156	720982-OA	. . COVER, BAND UOC: YNN, 86R	1
6	PAFZZ	PAOZZ	5305-00-068-0500	96906	MS90725-3	. . SCREW, CAP, HEX HEAD UOC: YNN, 86R	28
7	XDFZZ	XB		36156	718514-01	. . COVER, TOP UOC: YNN, 86R	1
8	XDFZZ	XB		36156	718516-01	. . COVER, LOUVERED, LS UOC: YNN, 86R	1
9	XDFZZ	XB		36156	71851501	. . COVER, LOUVERED, RS UOC: YNN, 86R	1
10	XDFZZ	XB		36156	718517-01	. . COVER, LOUVERED, BO UOC: YNN, 86R	1
11	XDFZZ	XB		30554	88-21804	. . PANEL, SIDE VENT UOC: YNN, 86R	1
12	PAFZZ	PAOZZ	5940-00-113-9820	96906	MS20659-128	. . TERMINAL, LUG USED ON STATOR LEADS UOC: YNN, 86R	10
13	PAFZZ	PAOZZ	5940-00-113-0954	96906	MS20659-165	. . TERMINAL, LUG USED ON EXCITER LEADS UOC: YNN, 86R	2
14	PAFZZ	PAOZZ	5305-00-068-0502	80205	MS90725-6	. . SCREW, CAP, HEX HEAD UOC: YNN, 86R	4
15	XDFZZ	XB		36156	752818-OA	. . LEAD CLAMP ASSEMBLY UOC: YNN, 86R	1
16	XDFZZ	XB		36156	834822-01	. . GASKET, LEAD CLAMP UOC: YNN, 86R	1
17	XDFZZ	XB		36156	718337-01	. . PANEL, LEAD EXIT UOC: YNN, 86R	1
18	PAFZZ	PAOZZ	5305-01-283-8664	80205	MS90725-110	. . SCREW, CAP, HEX HEAD UOC: YNN, 86R	4
19	PAFZZ	PAOZZ	5310-00-011-6121	96906	MS35338-67	. . WASHER, LOCK UOC: YNN, 86R	4
20	XDFFF	XB		36156	703513-01	. . ENDBELL UOC: YNN, 86R	1
21	PAFZA	PAFZA	5305-01-080-5721	88044	AN565C416H6	. . SETSCREW UOC: YNN, 86R	8
22	PBFZZ	PBOZZ	6115-01-385-3104	36156	789293-0A	. . STATOR, GENERATOR UOC: YNN, 86R	1
23	PAFZZ	PAOZZ	5330-01-369-7318	36156	865873-01	. . PACKING, PREFORMED UOC: YNN, 86R	1

(1) ITEM NO	(2) SMR CODE a. ARMY	(2) SMR CODE b. AIR FORCE	(3) NSN	(4) CAGEC	(5) PART NUMBER	(6) DESCRIPTION AND USABLE ON CODE (UOC)	(7) QTY
24	XDFFF	XB		36156	777068-0A	. . ROTOR ASSY 400 HZ (NOT SHOWN) UOC: YNN, 86R	1
25	PAFZZ	PAOZZ	3110-00-155-6298	21335	312KDD	. . . BEARING, BALL, ANNULA UOC: YNN, 86R	1
26	PAFZZ	PAOZZ	5305-00-068-0501	80205	MS90725-5	...SCREW, CAP, HEXHEAD UOC: YNN, 86R	4
27	PAFZZ	PAOZZ	5310-00-274-8715	80205	MS35338-63	...WASHER,LOCK UOC: YNN, 86R	4
28	PAFFF	PAFFF	2920-01-298-6321	36156	777056-0A	...RECTIFIER ASSEMBLY (NOT SHOWN) UOC: YNN, 86R	1
29	PAFZZ	PAOZZ	5310-00-761-6882	96906	MS51967-2	....NUT,PLAIN,HEX UOC: YNN, 86R	4
30	PAFZZ	PAOZZ	5310-00-550-1130	80205	MS35333-40	....WASHER,LOCK UOC: YNN, 86R	21
31	PAFZZ	PAOZZ	5305-00-068-0500	96906	MS90725-3	....SCREW, CAP, HEXHEAD UOC: YNN, 86R	9
32	PAFZZ	PAOZZ	5940-00-230-0515	96906	MS25036-154	....TERMINAL, LUG UOC: YNN, 86R	6
33	PAFZZ	PAOZZ	5310-01-494-2731	80205	AN3154	....NUT,PLAIN,HEX UOC: YNN, 86R	6
34	PAFZZ	PAOZZ	5961-01-067-9493	81349	JANTX1N1190R	....SEMICONDUCTOR DEVIC REVERSE, 35 AMP, 600 VOLT UOC: YNN, 86R	3
35	PAFZZ	PAOZZ	5961-00-154-7046	81349	JANTX1N1190	....SEMICONDUCTOR DEVICE, 35 AMP, 600 VOLT UOC: YNN, 86R	3
36	PAFZZ	PAOZZ	5305-00-984-6196	80205	MS35206-248	....SCREW, MACHINE UOC: YNN, 86R	1
37	PAFZZ	PAOZZ	5310-00-934-9757	80205	MS35649-282	....NUT,PLAIN,HEX UOC: YNN, 86R	1
38	PAFZZ	PAOZZ	5310-00-559-0070	96906	MS35333-38	....WASHER,LOCK UOC: YNN, 86R	1
39	PAFZZ	PAOZZ	5340-01-302-2960	96906	MS25281-4	....CLAMP, LOOP UOC: YNN, 86R	1
40	MFFZZ	MOO		30554	88-21649-19	....WIRE, ELECTRICAL MAKE FROM P/N 88-20444-2 (30554), 14 GA WIRE, AS REQUIRED UOC: YNN, 86R	6
41	MFFZZ	MOO		30554	88-21649-17	....INSULATION, SLEEVING MAKE FROM P/N M3190/03-17-0 (81349), AS REQUIRED UOC: YNN, 86R	2
42	PAFZZ	PAOZZ	5305-00-988-1727	80205	MS35206-283	....SCREW, MACHINE UOC: YNN, 86R	2
43	XDFZZ	XB		36156	B-718930-01	....PLATE,RECTIFIER MO UOC: YNN, 86R	3
44	XDFZZ	XB		36156	B-718817-01	....PLATE,RECTIFIER IN UOC: YNN, 86R	1
45	XDFZZ	XB	6130-01-533-2167	36156	707807-02	...RECTIFIER HUB UOC: YNN, 86R	1
46	XDFFF	XB	6115-01-533-2172	36156	791150-0A	...ROTOR ASSY, EXCITER UOC: YNN, 86R	1
47	XAFZZ	XA	5365-00-804-7654	96906	MS16624-1300	...RING, RETAINING UOC: YNN, 86R	1

(1) ITEM NO	(2) SMR CODE a. ARMY	b. AIR FORCE	(3) NSN	(4) CAGEC	(5) PART NUMBER	(6) DESCRIPTION AND USABLE ON CODE (UOC)	(7) QTY
48	PAFZZ	PAOZZ	5315-00-847-3531	96906	MS20066-356	. . . KEY, MACHINE UOC: YNN, 86R	1
49	PAFZZ	PAOZZ	5940-00-143-4777	96906	MS25036-157	...TERMINAL, LUG UOC: YNN, 86R	5
50	XDFZZ	XB		36156	707329-01	. . . HUB, DRIVE UOC: YNN, 86R	1
51	PAFZZ	PAOZZ	5315-00-227-6415	96906	MS20066-354	. . . KEY, MACHINE UOC: YNN, 86R	1
52	PBFZZ	PBOZZ	2930-01-329-4857	36156	716327-0A	...IMPELLER, FAN, AXIAL (NOT SHOWN) UOC: YNN, 86R	1
53	PAFZZ	PAOZZ	5310-00-984-3806	96906	MS51922-9	....NUT,SELF-LOCKING, HEX UOC: YNN, 86R	2
54	PAFZZ	PAOZZ	5305-00-225-8507	96906	MS90725-43	....SCREW, CAP, HEXHEAD UOC: YNN, 86R	2
55	XAFZZ	XA		30554	88-21783	....FAN,HALF UOC: YNN, 86R	2
56	PAFZZ	PAOZZ	5305-00-269-2801	80205	MS90726-58	. . SCREW, CAP, HEX HEAD UOC: YNN, 86R	6
57	PAFZZ	PAOZZ	5310-00-087-7493	96906	MS27183-13	. . WASHER, FLAT UOC: YNN, 86R	8
58	PAFZZ	PAOZZ	5305-00-269-3239	80204	B1821BH038F138N	. . SCREW, CAP, HEX HEAD UOC: YNN, 86R	2
59	XDFZZ	XB	5365-01-384-9169	30554	88-22339	. . SPACER, COUPLER UOC: YNN, 86R	2
60	XDFZZ	XB		36156	702806-01	. . DISC, COUPLING UOC: YNN, 86R	2
61	PAFZZ	PAOZZ	5305-00-282-3249	88044	AN565C816H16	. . SETSCREW UOC: YNN, 86R	2
62	XDFZZ	XB		36156	783622-0A	. . STATOR CORE, MAIN W UOC: YNN, 86R	1
63	PAFZZ	PAOZZ	5306-00-017-6143	96906	MS51937-7	. . BOLT, EYE UOC: YNN, 86R	1
64	PAFZZ	PAOZZ	5310-00-763-8920	96906	MS51967-20	. . NUT, PLAIN, HEX UOC: YNN, 86R	1
65	MDFZZ	MDO		36156	A-9696	. . LABEL, WARNING UOC: YNN, 86R	1
66	PAFZZ	PAOZZ	5305-00-253-5615	80205	MS21318-21	. . SCREW, DRIVE UOC: YNN, 86R	4
67	MDFZZ	MDO		30554	88-20064-6	. . PLATE, IDENTIFICATION UOC: YNN, 86R	1
68	XDFZZ	XB		30554	88-21798	. . FRAME, STATOR HOUSING UOC: YNN, 86R	1

END OF FIGURE

FIELD AND SUSTAINMENT MAINTENANCE

15 kW 50/60 AND 400 Hz SKID MOUNTED TACTICAL QUIET GENERATOR SETS

GROUP 12 ENGINE ASSEMBLY: ENGINE INSTALLATION

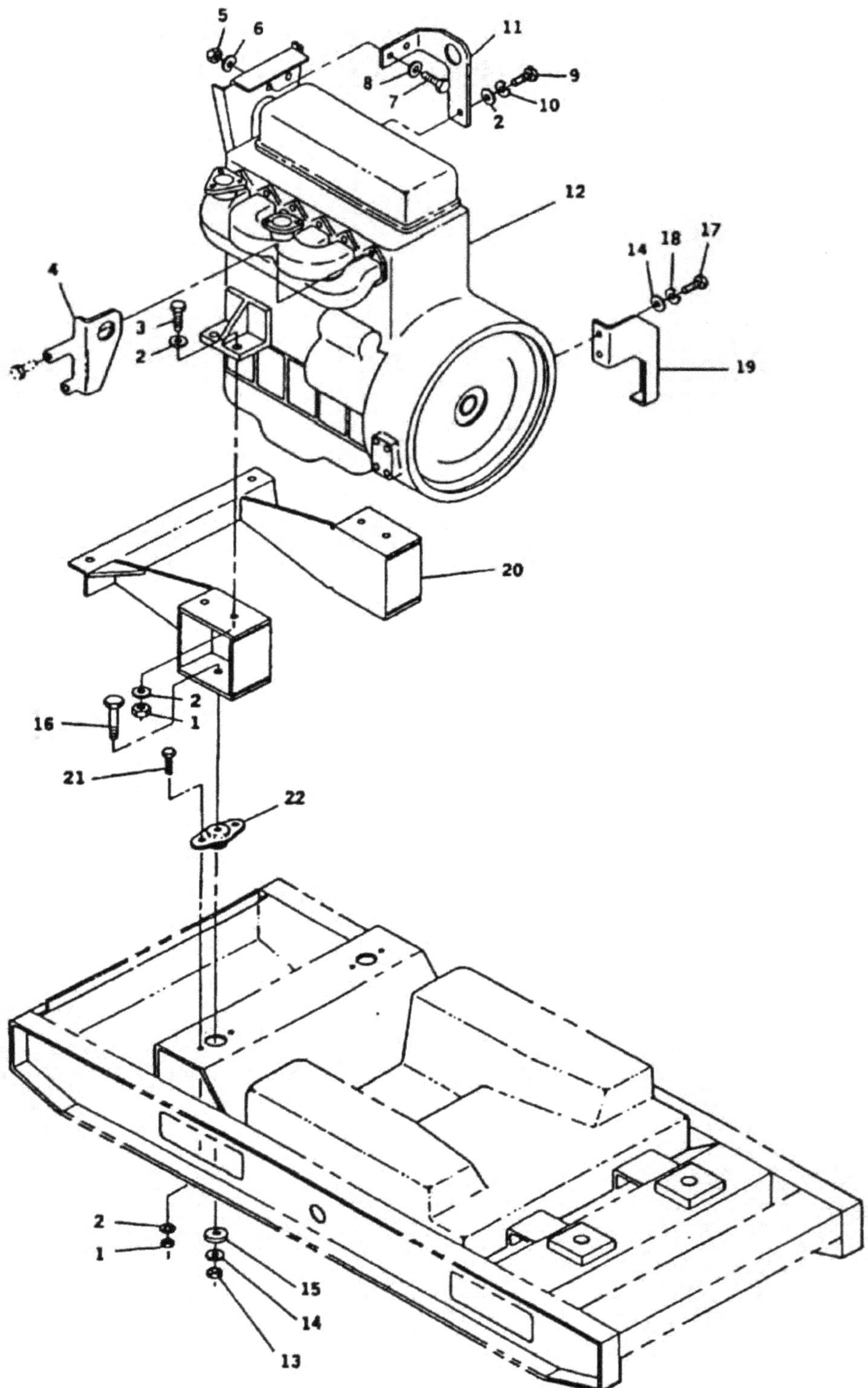

Figure 27. Engine Installation.

(1) ITEM NO	(2) SMR CODE a. ARMY	b. AIR FORCE	(3) NSN	(4) CAGEC	(5) PART NUMBER	(6) DESCRIPTION AND USABLE ON CODE (UOC)	(7) QTY
						GROUP 12 ENGINE ASSEMBLY	
						FIG. 27 ENGINE INSTALLATION	
1	PAFZZ	PAOZZ		96906	MS51493-5	. NUT, SELF-LOCKING UOC: EMK, YNN	8
2	PAFZZ	PAOZZ	5310-01-257-7590	96906	MS51412-7	. WASHER, FLAT UOC: EMK, YNN	14
3	PAFZZ	PAOZZ	5305-00-725-2317	80204	B1821BH038C150N	. SCREW, CAP, HEX HEAD UOC: EMK, YNN	4
4	XDFZZ	XB		30554	88-22321	. BRACKET, ENGINE REAR UOC: EMK, YNN	1
5	PAFZZ	PAOZZ	5310-01-466-6312	30554	88-22790-1	. NUT, PLAIN, HEX UOC: EMK, YNN	2
6	PAFZZ	PAOZZ	5310-00-274-8715	80205	MS35338-63	. WASHER, LOCK UOC: EMK, YNN	2
7	PAFZZ	PAOZZ	5305-01-056-1501	30554	88-20260-32	. SCREW, HEX WASHERHEAD UOC: EMK, YNN	2
8	PAFZZ	PAOZZ	5310-01-103-6042	96906	MS51412-4	. WASHER, FLAT UOC: EMK, YNN	2
9	XDFZZ	XB		80204	B18231B8025N	. SCREW, CAP, HEX HEAD UOC: EMK, YNN	1
10	PAFZZ	PAOZZ	5310-01-099-9532	30554	88-22331-1	. WASHER, LOCK UOC: EMK, YNN	1
11	XDFZZ	XB		30554	88-22425	. STRAP, ENGINE LIFTING UOC: EMK, YNN	1
12	PAFHH	PAFHH	2815-01-350-2207	30554	88-22315	. ENGINE, DIESEL (SEE TM 9-2815-254-24P FOR PARTS) UOC: EMK, YNN	1
13	PAFZZ	PAOZZ	5310-01-231-7459	96906	MS51943-9	. NUT, SELF-LOCKING UOC: EMK, YNN	2
14	PAFZZ	PAOZZ	5310-01-266-4641	96906	MS51412-9	. WASHER, FLAT UOC: EMK, YNN	8
15	PBFZZ	PBOZZ	5310-01-365-3190	30554	88-20247-2	. WASHER, FLAT UOC: EMK, YNN	2
16	PAFZZ	PAOZZ	5305-00-071-2075	80204	B1821BH050C300N	. SCREW, CAP, HEX HEAD UOC: EMK, YNN	2
17	PAFZZ	PAOZZ	5305-00-071-2066	80204	B1821BH050C100N	. SCREW, CAP, HEX HEAD UOC: EMK, YNN	4
18	PAFZZ	PAOZZ	5310-00-011-6121	96906	MS35338-67	. WASHER, LOCK UOC: EMK, YNN	4
19	XDFZZ	XB		30554	88-22135	. SUPPORT, ENGINE REAR UOC: EMK, YNN	2
20	XDFZZ	XB		30554	88-22174	. SUPPORT, ENGINE UOC: EMK, YNN	1
21	PAFZZ	PAOZZ	5305-00-068-0510	80204	B1821BH038C100N	. SCREW, CAP, HEX HEAD UOC: EMK, YNN	4
22	PAFZZ	PAOZZ	5342-01-467-4366	30554	88-21071-3	. MOUNT, ENGINE UOC: EMK, YNN	2
						END OF FIGURE	

FIELD AND SUSTAINMENT MAINTENANCE

15 kW 50/60 AND 400 Hz SKID MOUNTED TACTICAL QUIET GENERATOR SETS

GROUP 12 ENGINE ASSEMBLY: ENGINE INSTALLATION

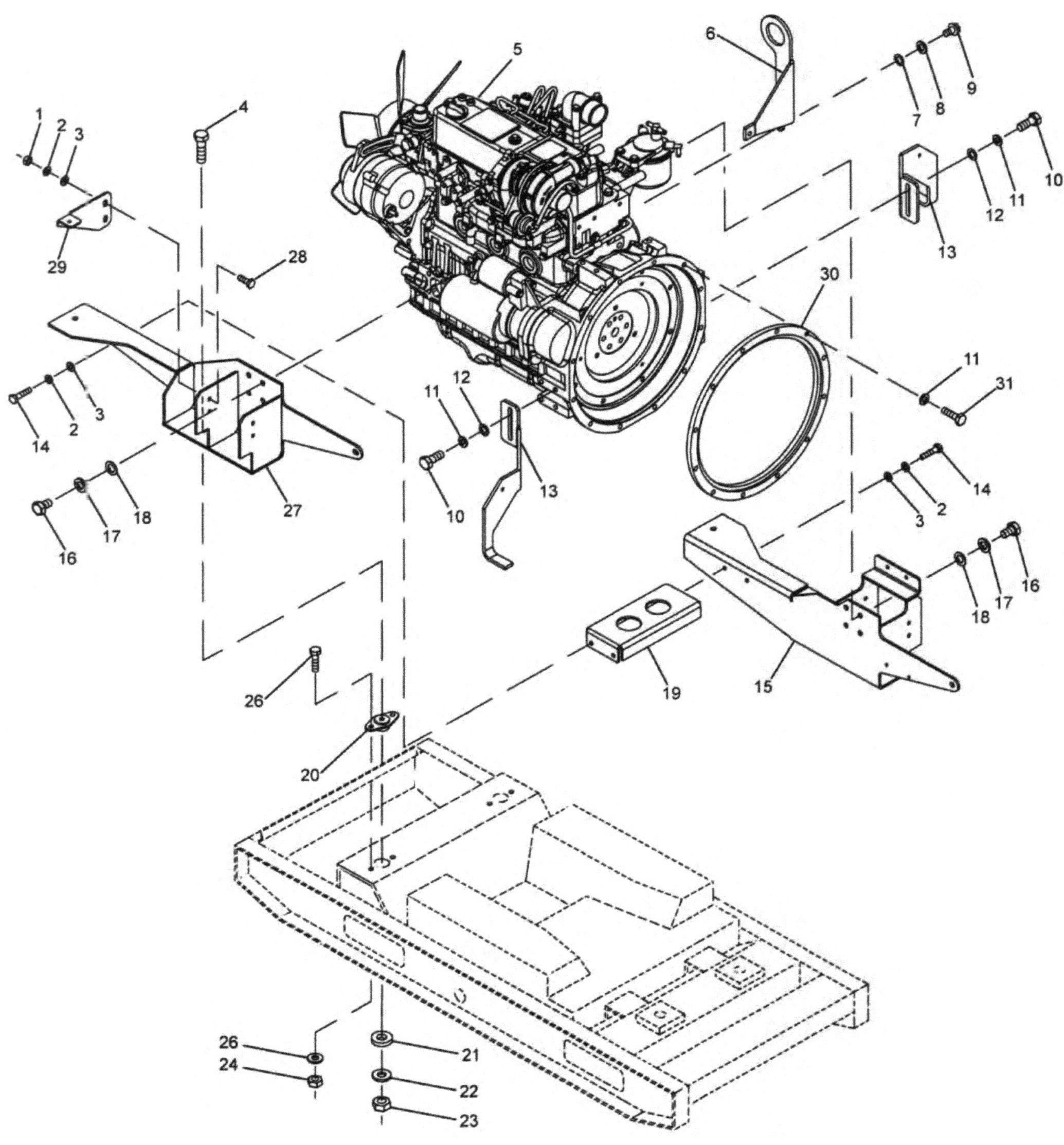

Figure 28. Engine Installation.

(1) ITEM NO	(2) SMR CODE a. ARMY	b. AIR FORCE	(3) NSN	(4) CAGEC	(5) PART NUMBER	(6) DESCRIPTION AND USABLE ON CODE (UOC)	(7) QTY
						GROUP 12 ENGINE ASSEMBLY	
						FIG. 28 ENGINE INSTALLATION	
1	PAFZZ	PAFZZ	5310-01-466-6312	30554	88-22790-1	. NUT, HEX HEAD UOC: 86Q, 86R	2
2	PAFZZ	PAFZZ		30554	88-20556-38	. WASHER, LOCK UOC: 86Q, 86R	6
3	PAFZZ	PAFZZ		30554	88-20033-21A	. WASHER, FLAT UOC: 86Q, 86R	6
4	PAFZZ	PAFZZ	5305-00-071-2075	80204	B1821BH050C300N	. SCREW, CAP, HEX HEAD UOC: 86Q, 86R	2
5	PAFHH	PAFHH		30554	97-24010	. DIESEL ENGINE (SEE TM 9-2815-538-24&P FOR PARTS) UOC: 86Q, 86R	1
6	XBFZZ	XBFZZ		30554	97-24023	. RING, LIFT, REAR UOC: 86Q, 86R	1
7	PAFZZ	PAFZZ	5310-01-531-4937	30554	88-20033-24A	. WASHER, FLAT UOC: 86Q, 86R	3
8	PAFZZ	PAFZZ	5310-01-099-9532	30554	88-22331-1	. WASHER, LOCK SPRING UOC: 86Q, 86R	3
9	PAFZZ	PAFZZ		81343	B18231A03014	. SCREW UOC: 86Q, 86R	3
10	PAFZZ	PAFZZ	5306-01-179-1438	80204	B18231A10025N	. SCREW, CAP, HEX HEAD UOC: 86Q, 86R	4
11	PAFZZ	PAFZZ	5310-01-368-3048	30554	88-22331-2	. WASHER, LOCK-SPRING UOC: 86Q, 86R	16
12	PAFZZ	PAFZZ	5310-01-532-0321	30554	88-20033-31A	. WASHER, FLAT UOC: 86Q, 86R	4
13	XBFZZ	XBFZZ		30554	97-24038	. SUPPORT, ENGINE, REAR UOC: 86Q, 86R	2
14	PAFZZ	PAFZZ	5305-01-379-5734	30554	88-20260-35	. SCREW/WASHER UOC: 86Q, 86R	4
15	XBFZZ	XBFZZ		30554	97-24013	. SUPPORT, RADIATOR, RH UOC: 86Q, 86R	1
16	PAFZZ	PAFZZ	5306-01-174-8738	80204	B18231A10020N	. BOLT, MACHINE UOC: 86Q, 86R	10
17	PAFZZ	PAFZZ	5310-01-368-3048	30554	88-22331-2	. WASHER, LOCK UOC: 86Q, 86R	10
18	PAFZZ	PAFZZ	5310-01-532-0321	30554	88-20033-31A	. WASHER, FLAT UOC: 86Q, 86R	10
19	XBFZZ	XBFZZ		30554	97-24014	. CROSS MEMBER UOC: 86Q, 86R	1
20	PAFZZ	PAFZZ	5342-01-467-4366	30554	88-21071-3	. MOUNT, ENGINE UOC: 86Q, 86R	2
21	PAFZZ	PAFZZ	5310-01-365-3190	30554	88-20247-2	. WASHER, SNUBBING UOC: 86Q, 86R	2
22	PAFZZ	PAFZZ	5310-01-266-4641	30554	88-20033-31A	. WASHER, FLAT UOC: 86Q, 86R	2
23	PAFZZ	PAFZZ	5310-01-470-2776	30554	88-20568-5	. NUT, SELF-LOCKING UOC: 86Q, 86R	2
24	PAFZZ	PAFZZ	5310-01-470-6529	30554	88-20568-4	. NUT, SELF-LOCKING UOC: 86Q, 86R	4
25	PAFZZ	PAFZZ		30554	88-20033-34A	. WASHER, FLAT UOC: 86Q, 86R	4

(1) ITEM NO	(2) SMR CODE a. ARMY	b. AIR FORCE	(3) NSN	(4) CAGEC	(5) PART NUMBER	(6) DESCRIPTION AND USABLE ON CODE (UOC)	(7) QTY
26	PAFZZ	PAFZZ	5305-00-068-0510	80204	B1821BH038C100N	. SCREW, CAP, HEX HEAD UOC: 86Q, 86R	4
27	XBFZZ	XBFZZ		30554	97-24012	. SUPPORT, RADIATOR, LH UOC: 86Q, 86R	1
28	PAFZZ	PAFZZ	5305-01-369-2166	30554	88-20260-30	. SCREW, CAP, HEX HEAD UOC: 86Q, 86R	2
29	XBFZZ	XBFZZ		30554	97-24039	. SUPPORT, GUARD REAR L UOC: 86Q, 86R	1
30	PBFZZ	PBFZZ		30554	97-24037	. SPACER, GENERATOR UOC: 86Q, 86R	1
31	PAFZZ	PAFZZ		30554	97-24116-5	. SCREW, CAP, HEX HEAD UOC: 86Q	12
31	PAFZZ	PAFZZ		30554	97-24116-6	. SCREW, CAP, HEX HEAD UOC: 86R	12

END OF FIGURE

FIELD AND SUSTAINMENT MAINTENANCE

15 kW 50/60 AND 400 Hz SKID MOUNTED TACTICAL QUIET GENERATOR SETS

GROUP 14 SKID BASE

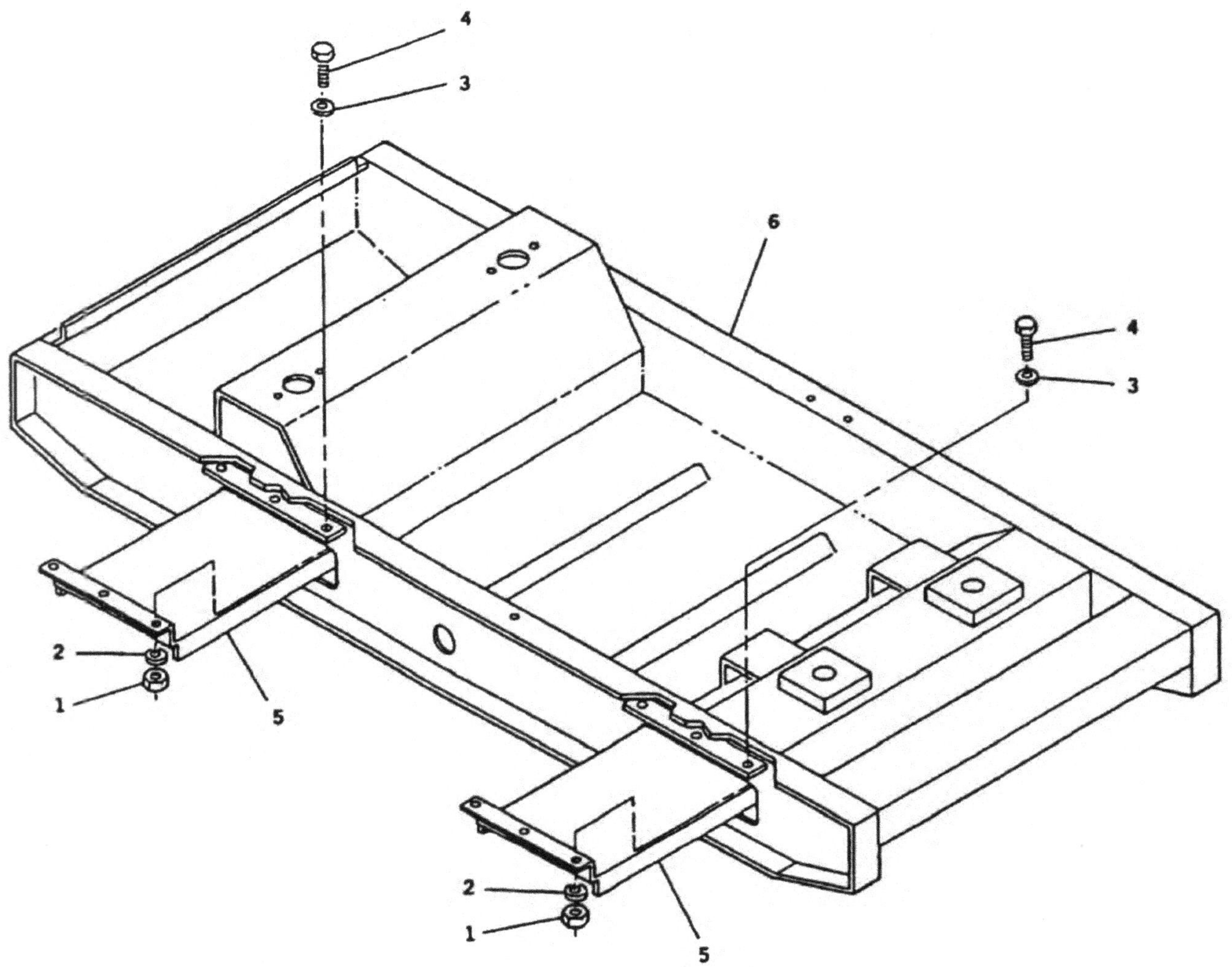

Figure 29. Skid Base.

(1)	(2) SMR CODE a.	b.	(3)	(4)	(5)	(6)	(7)
ITEM NO	ARMY	AIR FORCE	NSN	CAGEC	PART NUMBER	DESCRIPTION AND USABLE ON CODE (UOC)	QTY
						GROUP 14 SKID BASE	
						FIG. 29 SKID BASE	
1	PAFZZ	PAOZZ	5340-00-066-1235	30554	88-22790-3	. NUT, PLAIN, HEX	12
2	PAFZZ	PAOZZ	5310-00-011-5093	80205	MS35338-65	. WASHER, LOCK	12
3	PAFZZ	PAOZZ	5310-01-257-7590	96906	MS51412-7	. WASHER, FLAT	12
4	PAFZZ	PAOZZ	5305-00-068-0510	80204	B1821BH038C100N	. SCREW, CAP, HEX HEAD	12
5	XDFZZ	XB		30554	88-21731	. GUIDE, FORK LIFT	2
6	XDFZZ	XB		30554	88-22007	. SKID BASE ASSY	1

END OF FIGURE

FIELD AND SUSTAINMENT MAINTENANCE

15 kW 50/60 AND 400 Hz SKID MOUNTED TACTICAL QUIET GENERATOR SETS

GROUP 14 WINTERIZATION KIT

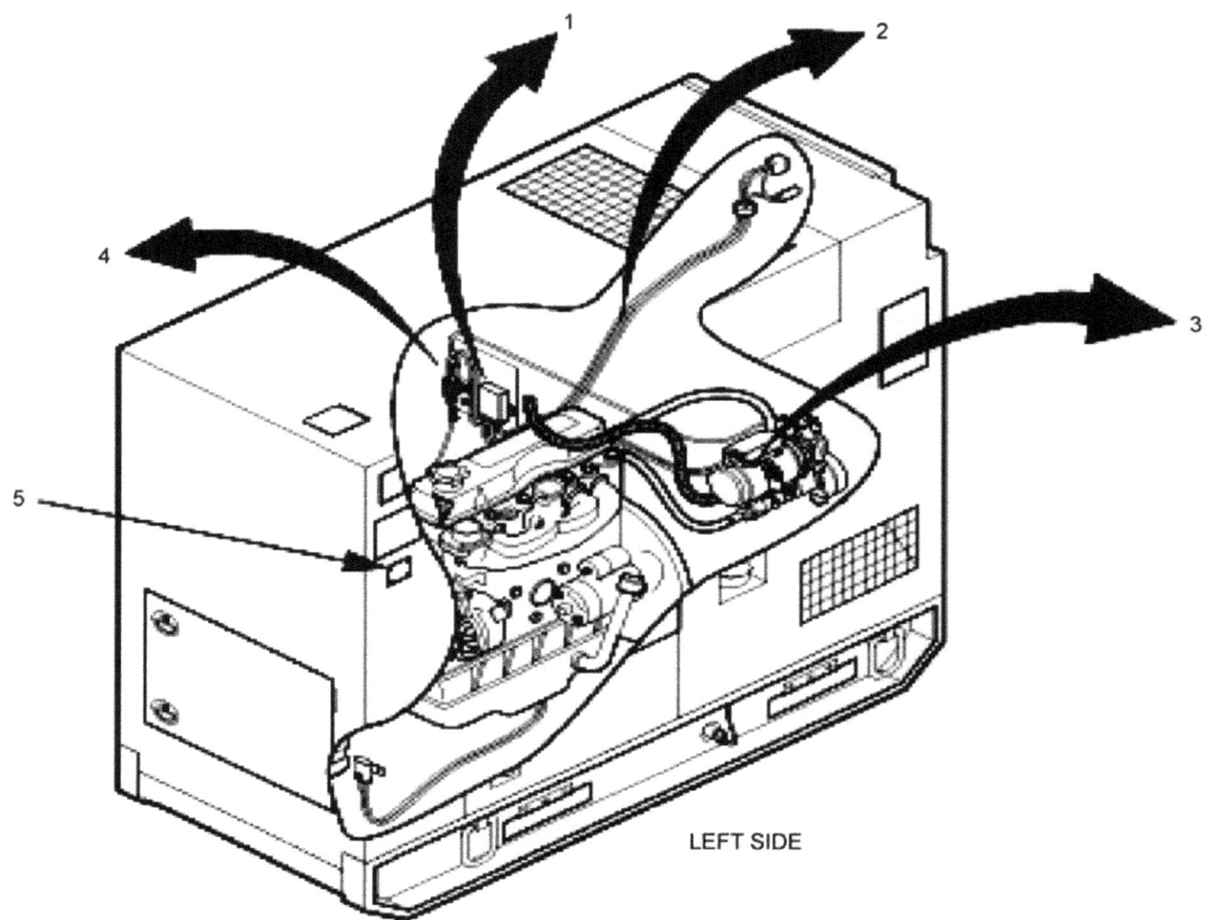

Figure 30. Winterization Kit.

(1) ITEM NO	(2) SMR CODE a. ARMY	b. AIR FORCE	(3) NSN	(4) CAGEC	(5) PART NUMBER	(6) DESCRIPTION AND USABLE ON CODE (UOC)	(7) QTY
						GROUP 14 WINTERIZATION KIT	
						FIG. 30 WINTERIZATION KIT	
1	PAFZZ	PAFZZ	2815-01-482-8924	30554	98-2014	. CONTROL UNIT SEE FIGURE 31 FOR PARTS	1
2	PAFFF	PAFFF	6150-01-496-7717	30554	98-2018-1	UOC: EMK, YNN .WIRING HARNESS, J1 SEE FIGURE 33 FOR PARTS	1
3	PAFFF	PAFFF	2990-01-483-4051	30554	98-2000	UOC: EMK, YNN . HEATER, COOLANT, ENG SEE FIGURE 34 FOR PARTS	1
4	PAFFF	PAFFF	2910-01-483-4069	30554	98-2013	UOC: EMK, YNN .PUMP, FUEL, ELECTRICAL AND ATTACHMENTS SEE FIGURE 33 FOR PARTS	1
5	XBFZZ	XBFZZ		30554	98-2053-3	UOC: EMK, YNN .PLATE IDENTIFICATION	1
6	PAFZZ	PAFZZ	6115-01-477-0566	30554	98-2015	UOC: EMK, YNN KIT, HEATER UOC, EMK, YNN (NOT SHOWN)	- -

END OF FIGURE

FIELD AND SUSTAINMENT MAINTENANCE

15 kW 50/60 AND 400 Hz SKID MOUNTED TACTICAL QUIET GENERATOR SETS

GROUP 14 WINTERIZATION KIT: CONTROL UNIT

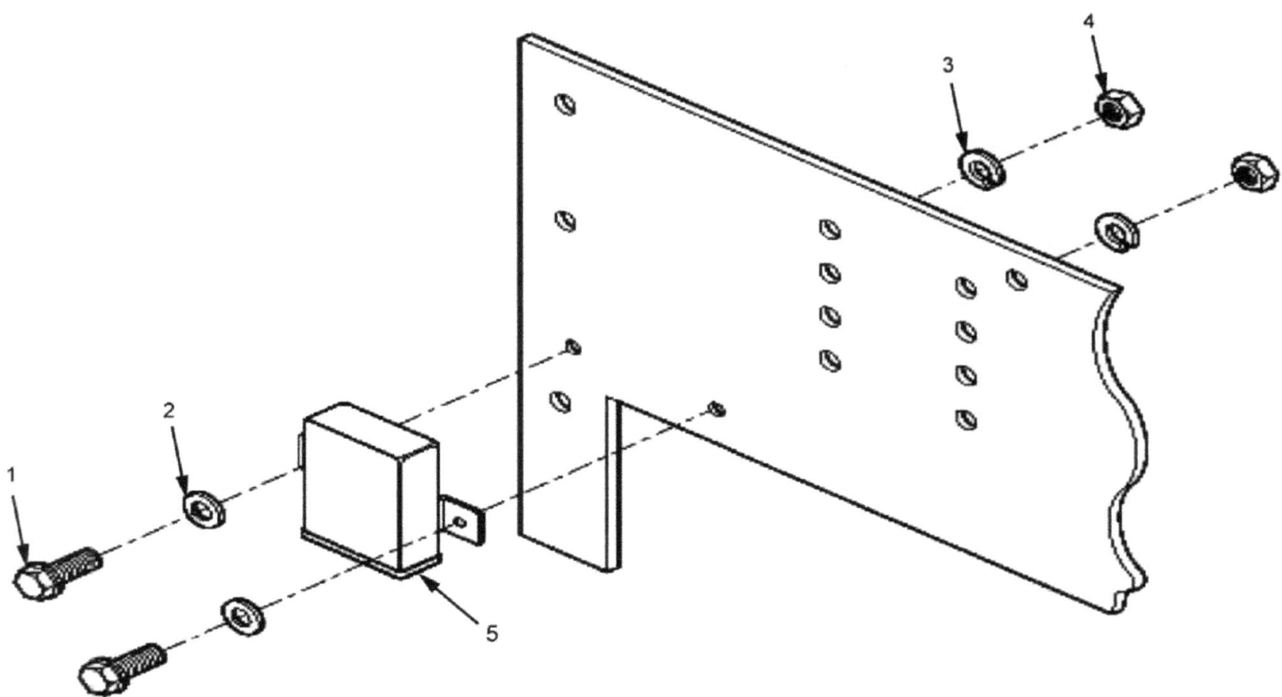

Figure 31. Control Unit.

(1) ITEM NO	(2) SMR CODE a. ARMY	b. AIR FORCE	(3) NSN	(4) CAGEC	(5) PART NUMBER	(6) DESCRIPTION AND USABLE ON CODE (UOC)	(7) QTY
						GROUP 14 WINTERIZATION KIT	
						FIG. 31 CONTROL UNIT	
1	XBFZZ	XBFZZ	5306-01-156-7663	30554	88-20260-21	. . SCREW, HEX WASHERHE UOC: EMK, YNN	2
2	XBFZZ	XBFZZ	5310-01-531-6085	30554	88-20033-10A	. . WASHER, FLAT UOC: EMK, YNN	2
3	XBFZZ	XBFZZ	5310-01-532-0291	30554	88-20556-4	. . WASHER, LOCK UOC: EMK, YNN	2
4	XBFZZ	XBFZZ	5310-00-533-2743	30554	13218E0320-49	. . NUT, PLAIN, HEX UOC: EMK, YNN	2
5	PAFZZ	PAFZZ	2915-01-482-8924	30554	98-2014	. CONTROL UNIT UOC: EMK, YNN	REF

END OF FIGURE

FIELD AND SUSTAINMENT MAINTENANCE

15 kW 50/60 AND 400 Hz SKID MOUNTED TACTICAL QUIET GENERATOR SETS

GROUP 14 WINTERIZATION KIT: OPERATING PLATES AND SWITCH

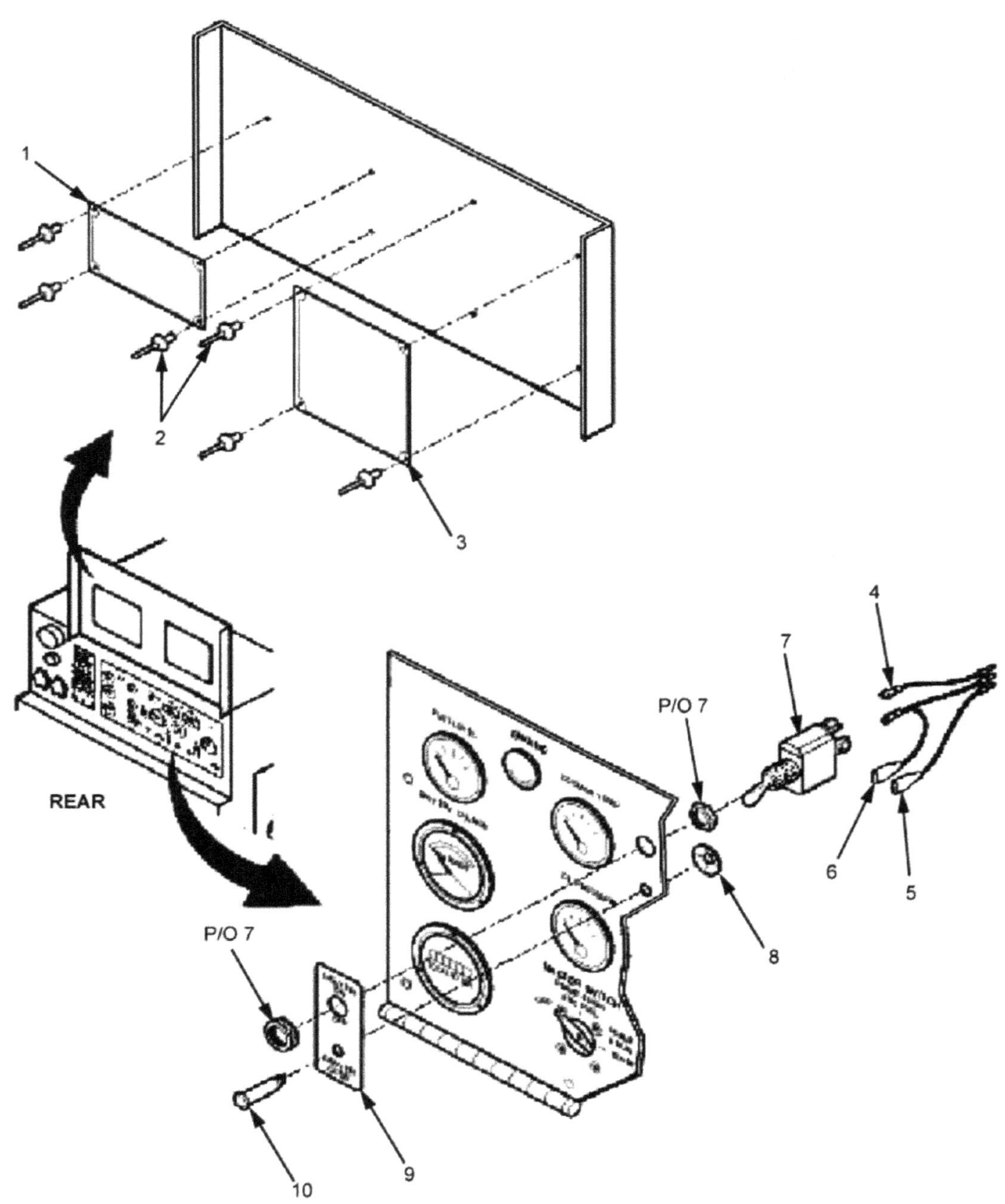

Figure 32. Operating Plates and Switch.

(1) ITEM NO	(2) SMR CODE a. ARMY	b. AIR FORCE	(3) NSN	(4) CAGEC	(5) PART NUMBER	(6) DESCRIPTION AND USABLE ON CODE (UOC)	(7) QTY
						GROUP 14 WINTERIZATION KIT	
						FIG. 32 OPERATING PLATES AND SWITCH	
1	MFFZZ	MOFZZ		30554	98-2041	. PLATE, OPERATING INST UOC: EMK, YNN	1
2	XBFZZ	XBFZZ	5320-00-932-1972	97403	13214E3789-2	. . RIVET, BLIND UOC: EMK, YNN	8
3	MFFZZ	MOFZZ		30554	98-2001	. PLATE, HEATER FUNCTION UOC: EMK, YNN	1
4	MFFFF	MOFFF		30554	98-2049-4	. LEAD, ELECTRICAL SEE FIGURE 39 FOR PARTS UOC: EMK, YNN	1
5	MFFFF	MOFFF		30554	98-2049-3	. LEAD, ELECTRICAL SEE FIGURE 39 FOR PARTS UOC: EMK, YNN	1
6	MFFFF	MOFFF		30554	98-2050	. LEAD, ELECTRICAL, SEE FIGURE 39 FOR PARTS UOC: EMK, YNN	1
7	XBFZZ	XBFZZ	5930-01-366-0048	30554	88-22141	. SWITCH, TOGGLE UOC: EMK, YNN	1
8	XBFZZ	XBFZZ		30554	98-2043	. . PUSH ON NUT UOC: EMK, YNN	1
9	XBFZZ	XBFZZ		30554	98-2021	. LABEL, HEATER SWITCH UOC: EMK, YNN	1
10	XBFZZ	XBFZZ		30554	98-2003	. LIGHT, INDICATOR UOC: EMK, YNN	1
						END OF FIGURE	

FIELD AND SUSTAINMENT MAINTENANCE

15 kW 50/60 AND 400 Hz SKID MOUNTED TACTICAL QUIET GENERATOR SETS
GROUP 14 WINTERIZATION KIT: PUMP, FUEL, ELECTRICAL AND ATTACHMENTS

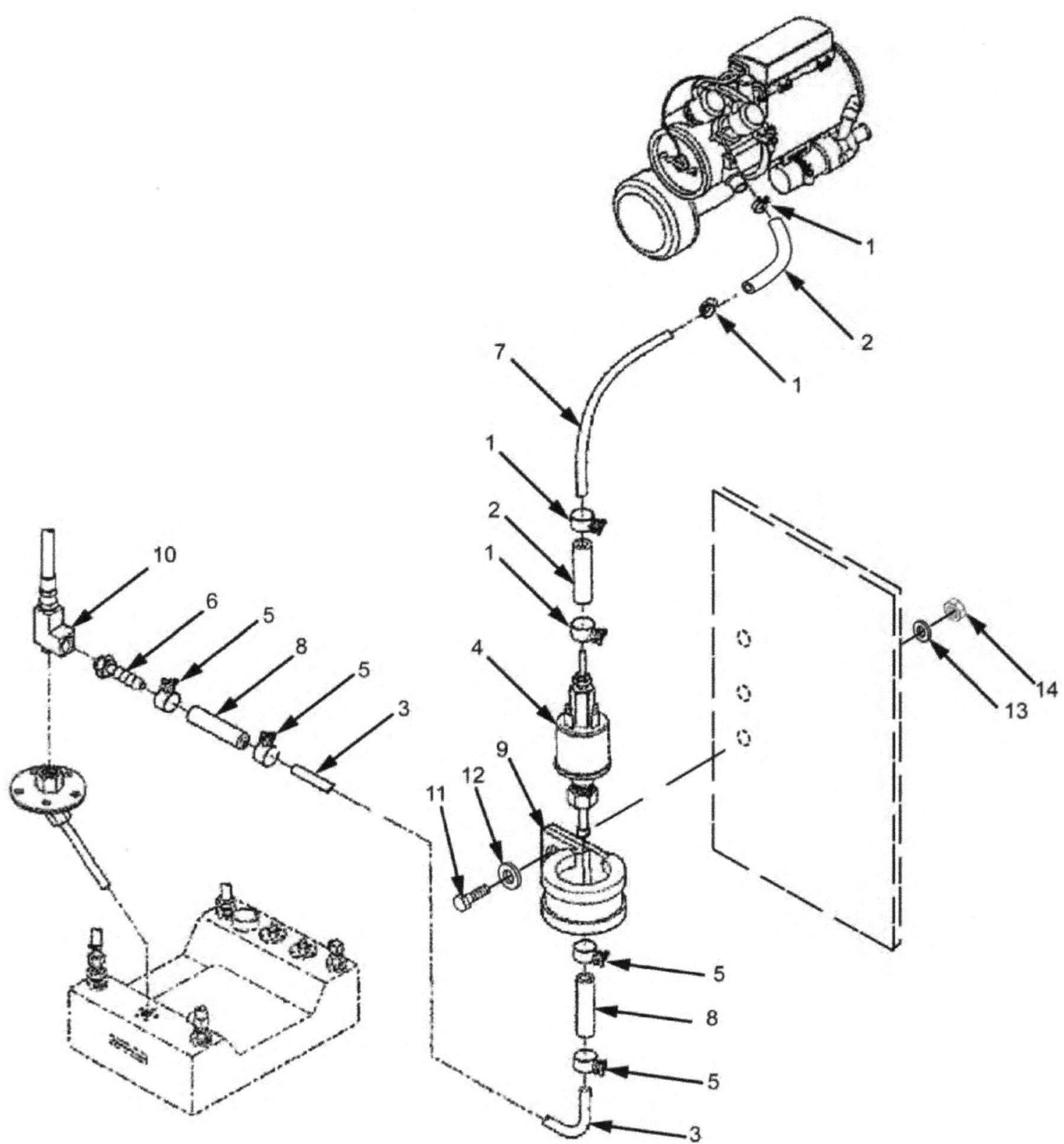

Figure 33. Pump, Fuel, Electrical and Attachments.

(1) ITEM NO	(2) SMR CODE a. ARMY	b. AIR FORCE	(3) NSN	(4) CAGEC	(5) PART NUMBER	(6) DESCRIPTION AND USABLE ON CODE (UOC)	(7) QTY
						GROUP 14 WINTERIZATION KIT	
						FIG. 33 PUMP, FUEL, ELECTRICAL AND ATTACHMENTS	
1	XBFZZ	XBFZZ		81343	J1508-10 TYPE D	. CLAMP, HOSE UOC: EMK, YNN	4
2	XBFZZ	XBFZZ		81343	J30R7-TYPE 1-5/32ID	. HOSE, NONMETALLIC IS 2 INCHES LONG UOC: EMK, YNN	2
3	PAFZZ	PAFZZ		1C645	DIN73378-6x2-PA12-HL-sw	. TUBING, 6X2MM, BLACK, TUBING IS 24 IN. L UOC: EMK, YNN	1
4	PAFZZ	PAFZZ	2910-01-483-4069	30554	98-2013	. PUMP FUEL UOC: EMK, YNN	REF
5	XBFZZ	XBFZZ		81343	J1508-13 TYPE D	. CLAMP, HOSE UOC: EMK, YNN	4
6	XBFZZ	XBFZZ		30554	98-2025-3	. ADAPTER, STRAIGHT, PI UOC: EMK, YNN	1
7	PAFZZ	PAFZZ		1C645	DIN73378-4x1.25-PA12-HL-nf	. TUBING, 4X1.25MM, NAT, TUBING IS UOC: EMK, YNN	1
8	XBFZZ	XBFZZ	4720-01-293-4415	81343	J30R7-TYPE 1-3/16ID	. HOSE NONMETALLIC HOSE IS 2 INCH LONG UOC: EMK, YNN	2
9	XBFZZ	XBFZZ		30554	88-20037WDG21	. CLAMP, LOOP, CUSHION UOC: EMK, YNN	1
10	XBFZZ	XBFZZ	4730-01-470-6199	81343	4-4-4-140424B	. TEE, PIPE UOC: EMK, YNN	1
11	XBFZZ	XBFZZ	5306-01-156-7663	30554	88-20260-21	. . SCREW, HEX WASHERHE UOC: EMK, YNN	1
12	XBFZZ	XBFZZ	5310-01-531-6085	30554	88-20033-10A	. . WASHER, FLAT UOC: EMK, YNN	1
13	XBFZZ	XBFZZ	5310-01-532-0291	30554	88-20556-4	. . WASHER, LOCK UOC: EMK, YNN	1
14	XBFZZ	XBFZZ	5310-00-533-2743	97403	13218E0320-49	. . NUT, PLAIN, HEX UOC: EMK, YNN	1

END OF FIGURE

FIELD AND SUSTAINMENT MAINTENANCE

15 kW 50/60 AND 400 Hz SKID MOUNTED TACTICAL QUIET GENERATOR SETS

GROUP 14 WINTERIZATION KIT: HEATER ASSEMBLY

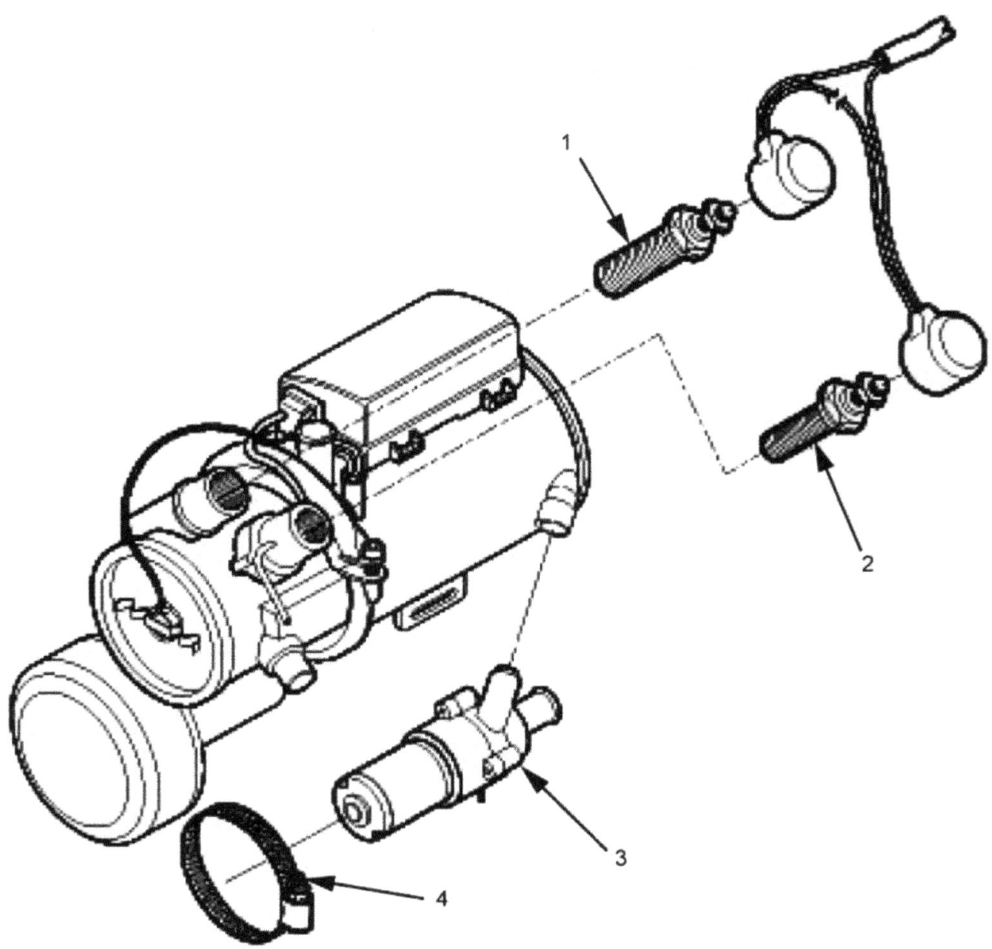

Figure 34. Heater Assembly.

(1) ITEM NO	(2) SMR CODE a. ARMY	b. AIR FORCE	(3) NSN	(4) CAGEC	(5) PART NUMBER	(6) DESCRIPTION AND USABLE ON CODE (UOC)	(7) QTY
						GROUP 14 WINTERIZATION KIT	
						FIG. 34 HEATER ASSEMBLY	
1	PAFZZ	PAFZZ	2920-12-187-7756	D2212	0102124405	. . GLOW PLUG UOC: EMK, YNN	1
2	PAFZZ	PAFZZ	5905-14-485-6282	F1425	251667010001	. . RESISTOR UOC: EMK, YNN	1
3	PAFZZ	PAFZZ	4320-12-326-9400	D8435	251671250100	. . WATER PUMP UOC: EMK, YNN	1
4	XBFZZ	XBFZZ	4730-12-167-2104	D8435	102064032050	. . CLAMP, HOSE UOC: EMK, YNN	1
5	PAFZZ	PAFZZ	2990-01-483-4051	30554	98-2000	. HEATER, COOLANT, ENG (NOT SHOWN) UOC: EMK, YNN	REF

END OF FIGURE

FIELD AND SUSTAINMENT MAINTENANCE

15 kW 50/60 AND 400 Hz SKID MOUNTED TACTICAL QUIET GENERATOR SETS
GROUP 14 WINTERIZATION KIT: HEATER ASSEMBLY ATTACHMENTS

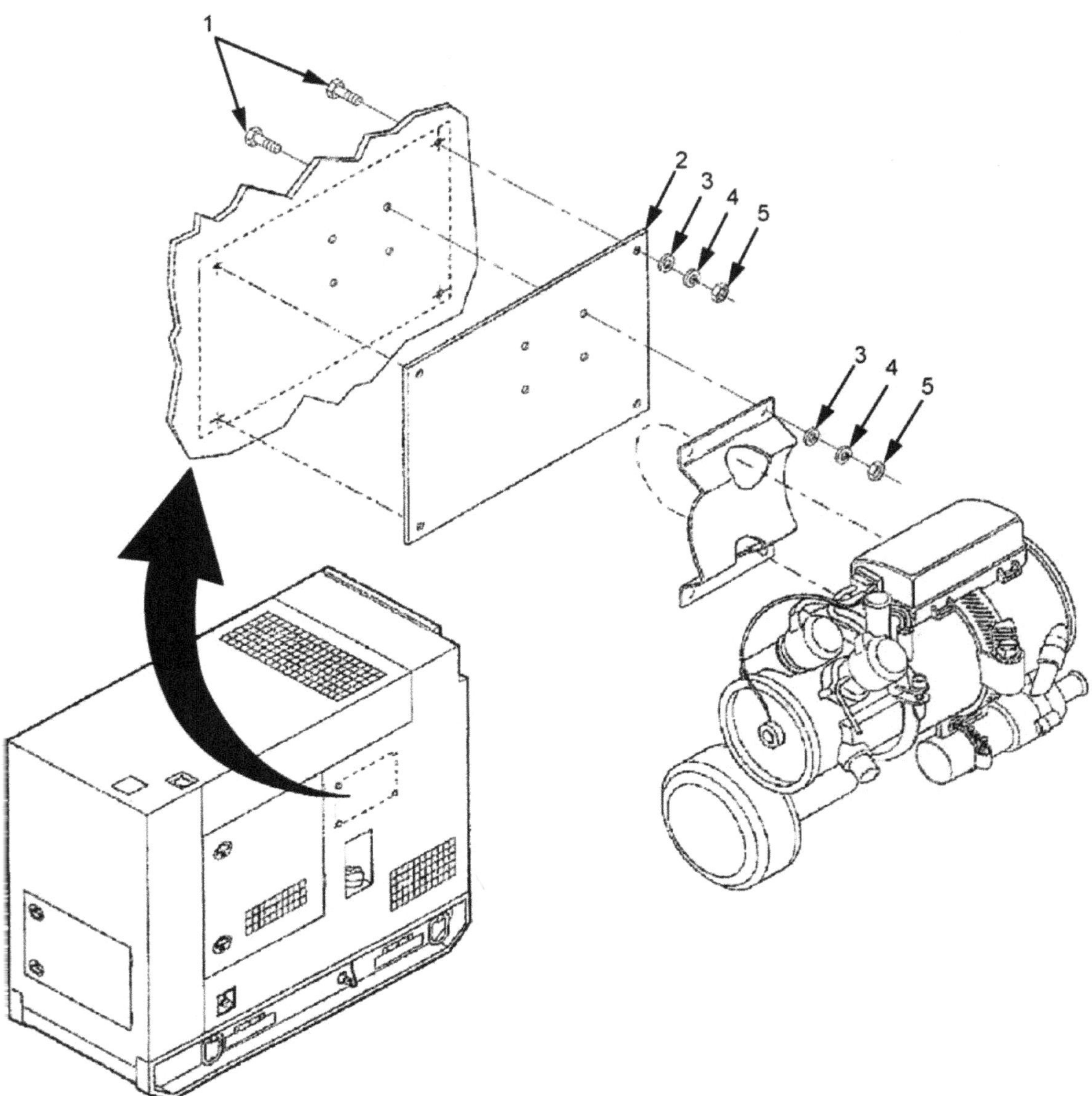

Figure 35. Heater Assembly Attachments.

(1) ITEM NO	(2) SMR CODE a. ARMY	b. AIR FORCE	(3) NSN	(4) CAGEC	(5) PART NUMBER	(6) DESCRIPTION AND USABLE ON CODE (UOC)	(7) QTY
						GROUP 14 WINTERIZATION KIT	
						FIG. 35 HEATER ASSEMBLY ATTACHMENTS	
1	XBFZZ	XBFZZ	5306-01-156-7663	20554	88-20260-21	. . SCREW, HEXWASHERHE UOC: EMK, YNN	8
2	XBFZZ	XBFZZ		30554	98-2002	. . PLATE, MOUNTING, HEAT UOC: EMK, YNN	1
3	XBFZZ	XBFZZ	5310-01-531-6085	30554	88-20033-10A	. . WASHER, FLAT UOC: EMK, YNN	8
4	XBFZZ	XBFZZ	55310-01-532-0291	30554	88-20556-4	. . WASHER, LOCK UOC: EMK, YNN	8
5	XBFZZ	XBFZZ	5310-00-533-2743	97403	13218E0320-49	. . NUT, PLAIN, HEX UOC: EMK, YNN	8

END OF FIGURE

FIELD AND SUSTAINMENT MAINTENANCE

15 kW 50/60 AND 400 Hz SKID MOUNTED TACTICAL QUIET GENERATOR SETS
GROUP 14 WINTERIZATION KIT: COOLANT HOSES AND ATTACHMENTS

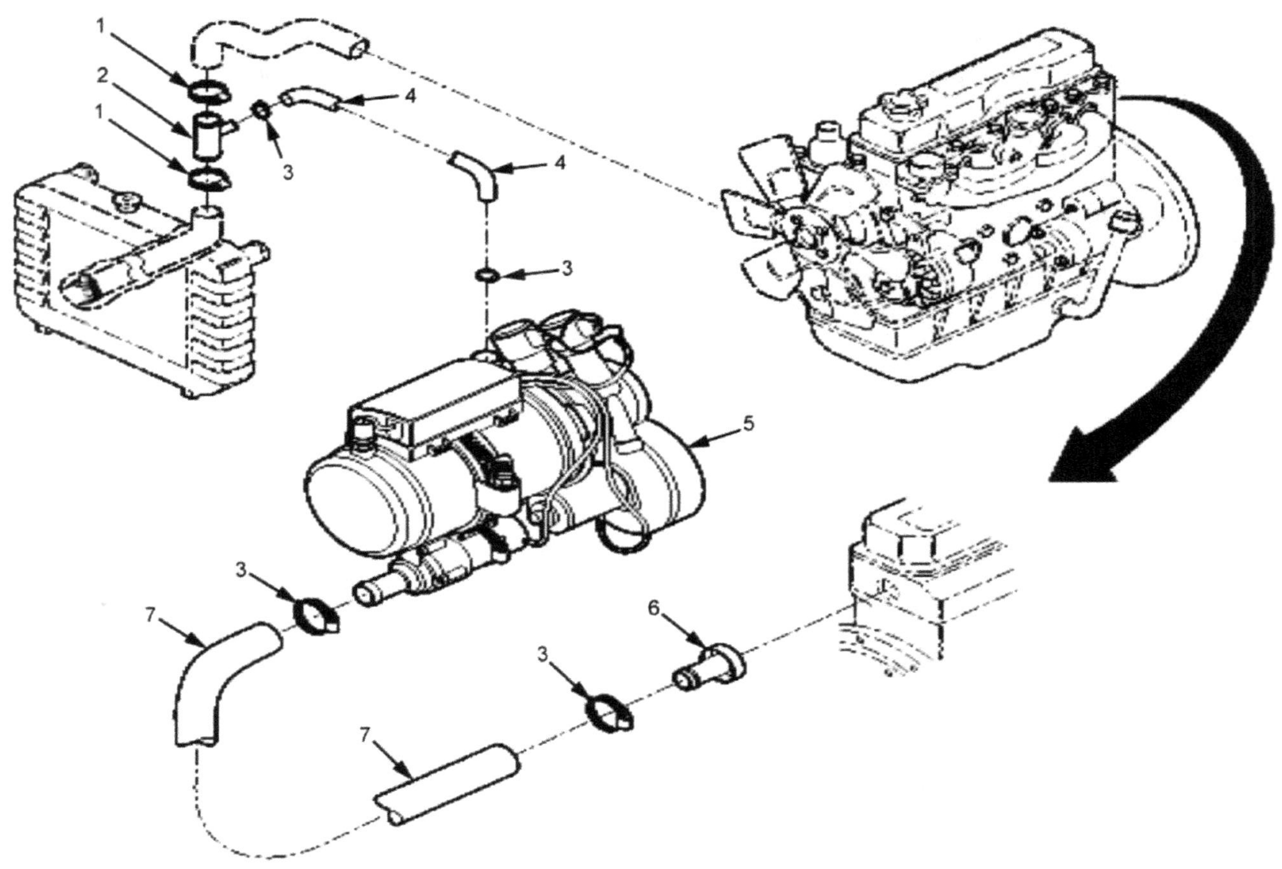

Figure 36. Coolant Hoses and Attachments.

(1) ITEM NO	(2) SMR CODE a. ARMY	b. AIR FORCE	(3) NSN	(4) CAGEC	(5) PART NUMBER	(6) DESCRIPTION AND USABLE ON CODE (UOC)	(7) QTY
						GROUP 14 WINTERIZATION KIT	
						FIG. 36 COOLANT HOSES AND ATTACHMENTS	
1	XBFZZ	XBFZZ	4730-00-908-3193	58536	J1670-24	. . CLAMP, TYPE F, SIZE 24 UOC: EMK, YNN	2
2	XBFZZ	XBFZZ		30554	98-2009	. . TEE UOC: EMK, YNN	1
3	XBFZZ	XBFZZ		81343	J1670-10	. . CLAMP, TYPE F, SIZE 10 UOC: EMK, YNN	4
4	PAFZZ	PAFZZ	4720-01-484-6062	30554	98-2059-7	. . HOSE, PREFORMED, 90 D, 60 IN. L UOC: EMK, YNN	1
5	PAFFF	PAFFF	2990-01-483-4051	30554	98-2000	. HEATER, COOLANT, ENG UOC: EMK, YNN	REF
6	XBFZZ	XBFZZ		30554	98-2022	. . PLUG, EXPANSION UOC: EMK, YNN	1
7	PAFZZ	PAFZZ	4720-01-484-6004	30554	98-2059-4	. . HOSE, PREFORMED, 90 D, 24 IN. L UOC: EMK, YNN	1

END OF FIGURE

FIELD AND SUSTAINMENT MAINTENANCE

15 kW 50/60 AND 400 Hz SKID MOUNTED TACTICAL QUIET GENERATOR SETS
GROUP 14 WINTERIZATION KIT: EXHAUST HOSES AND HEATER BRACE

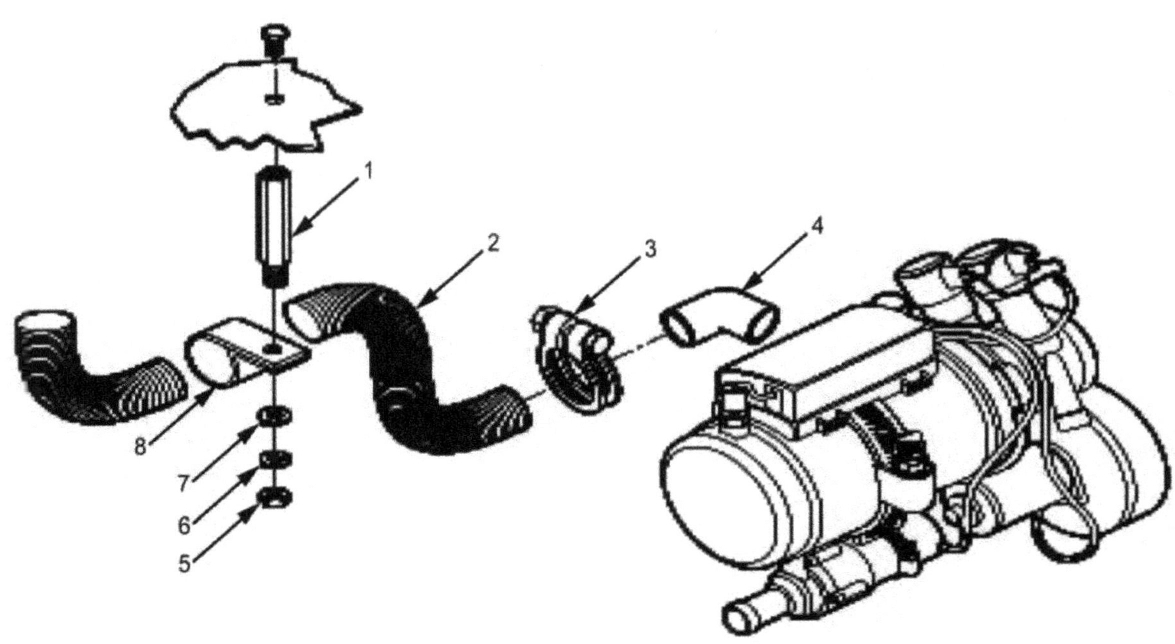

Figure 37. Exhaust Hoses and Heater Brace.

(1) ITEM NO	(2) SMR CODE a. ARMY	b. AIR FORCE	(3) NSN	(4) CAGEC	(5) PART NUMBER	(6) DESCRIPTION AND USABLE ON CODE (UOC)	(7) QTY
						GROUP 14 WINTERIZATION KIT	
						FIG. 37 EXHAUST HOSES AND HEATER BRACE	
1	XBFZZ	XBFZZ	5340-01-470-5184	30554	96-23615	. . STANDOFF, MALE-FEMALE UOC: EMK, YNN	1
2	XBFZZ	XBFZZ		30554	98-2019-2	. . HOSE METALLIC, 30 IN. L UOC: EMK, YNN	1
3	XBFZZ	XBFZZ		30554	98-2008-4	. . CLAMP, LOOP UOC: EMK, YNN	1
4	XBFZZ	XBFZZ	4730-00-235-4066	30554	98-2024-5	. . ELBOW, 90 DEGREE UOC: EMK, YNN	1
5	XBFZZ	XBFZZ	5310-00-533-2743	97403	13218E0320-49	. . NUT, PLAIN, HEX UOC: EMK, YNN	1
6	XBFZZ	XBFZZ	5310-01-532-0291	30554	88-20556-4	. . WASHER, LOCK UOC: EMK, YNN	1
7	XBFZZ	XBFZZ	5310-01-531-6085	30554	88-20033-10A	. . WASHER, FLAT UOC: EMK, YNN	1
8	XBFZZ	XBFZZ	5340-00-550-5943	30554	98-2017-3	. . CLAMP, LOOP UOC: EMK, YNN	1

END OF FIGURE

FIELD AND SUSTAINMENT MAINTENANCE

15 kW 50/60 AND 400 Hz SKID MOUNTED TACTICAL QUIET GENERATOR SETS

GROUP 14 WINTERIZATION KIT: WIRING HARNESS

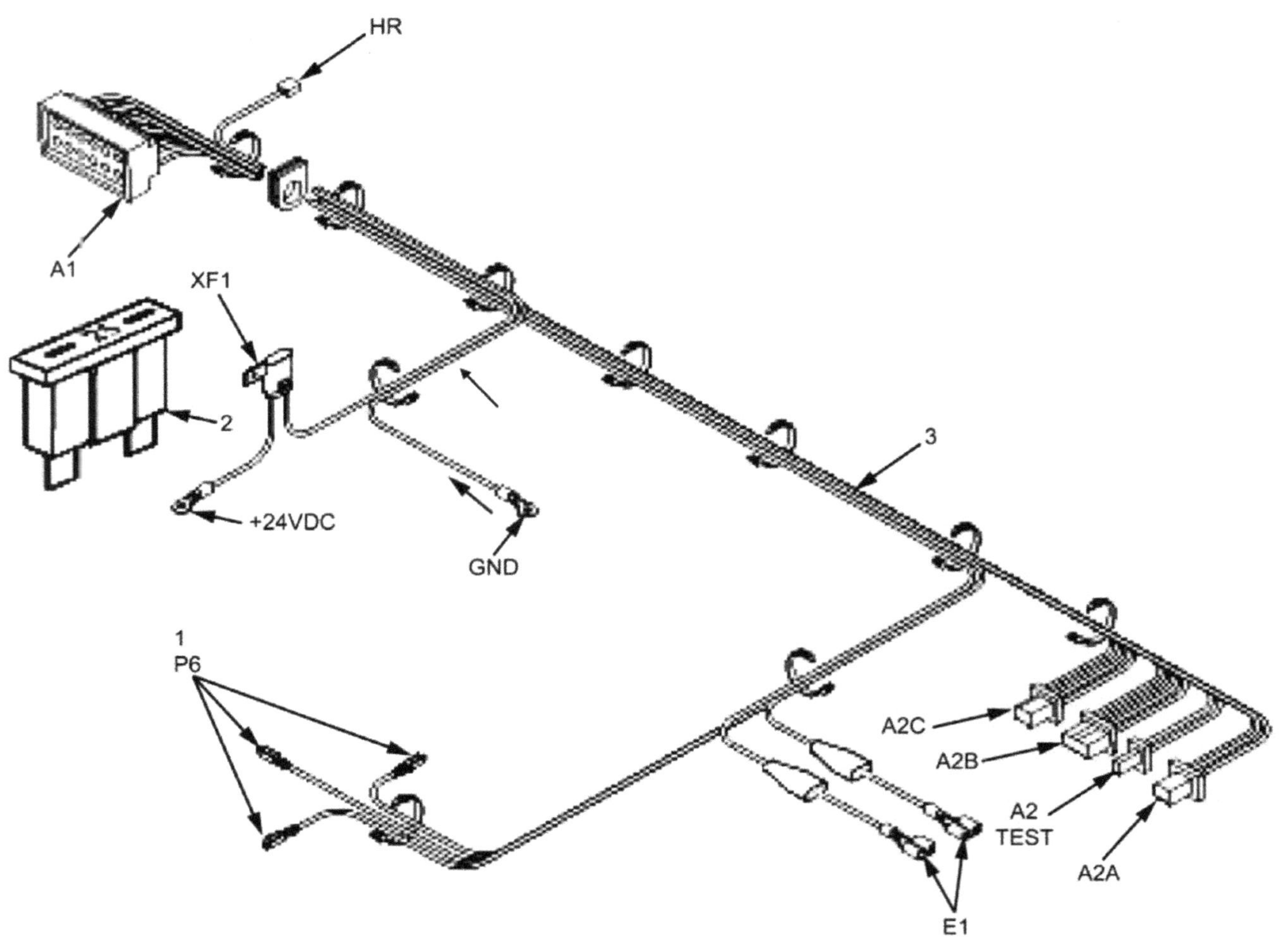

Figure 38. Wiring Harness.

(1)	(2) SMR CODE a.	b.	(3)	(4)	(5)	(6)	(7)
ITEM NO	ARMY	AIR FORCE	NSN	CAGEC	PART NUMBER	DESCRIPTION AND USABLE ON CODE (UOC)	QTY
						GROUP 14 WINTERIZATION KIT	
						FIG. 38 WIRING HARNESS	
1	XBFZZ	XBFZZ	5999-01-170-0558	30554	88-21942	. . CONNECTOR, P6 UOC: EMK, YNN	3
2	PAFZZ	PAFZZ	5920-01-414-6436	58536	AA55569/02-009	. . FUSE, INCLOSED, LINK UOC: EMK, YNN	1
3	PAFFF	PAFFF	6150-01-496-7717	30554	98-2018-1	. WIRING HARNESS UOC: EMK, YNN	REF
						END OF FIGURE	

FIELD AND SUSTAINMENT MAINTENANCE

15 kW 50/60 AND 400 Hz SKID MOUNTED TACTICAL QUIET GENERATOR SETS

GROUP 14 WINTERIZATION KIT: ELECTRICAL LEADS

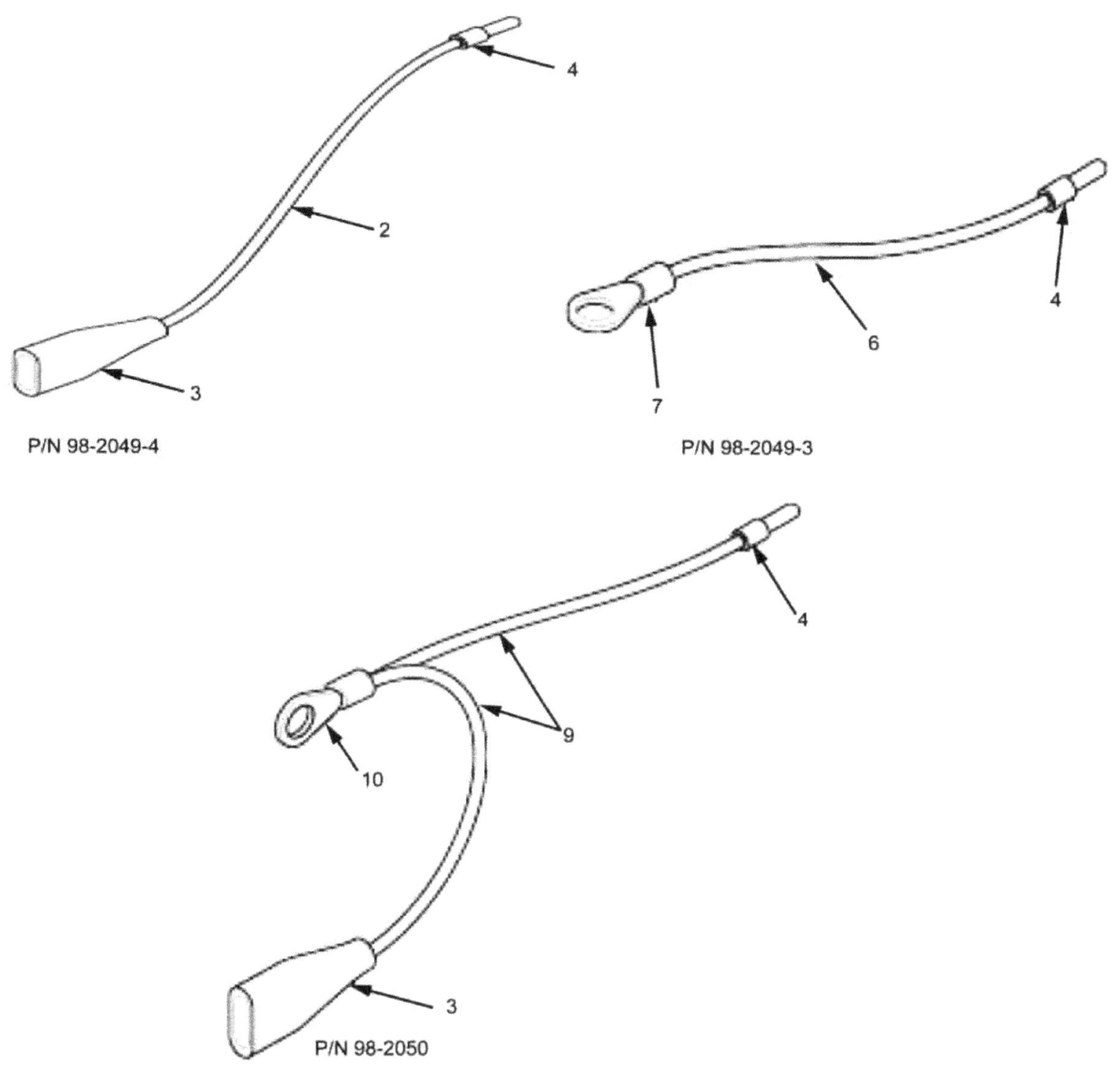

PART NUMBER	TERMINATION FROM	FIND #	TERMINATION TO	FIND#	WIRE FIND#	DIMENSION
98-2049-4	J6-17	4	DS1-	3	2	44"
98-2049-3	J6-16	4	S1-2	7	6	44"
98-2050	J6-24	4	S1-3	10	9	44"
	DS1+	3	S1-3	10	9	5"

Figure 39. Electrical Leads (Sheet 1 of 2).

WIRE NO.	TERMINATION FROM	ITEM NAME	TERMINATION TO	ITEM NAME	WIRE AWG.
1	A1-1	Heater Assembly	A2C-2	Control Unit	18
2	A1-2	Heater Assembly	GND	—	18
3	A1-3	Heater Assembly	GND	—	18
4	A1-4	Heater Assembly	A2C-5	Control Unit	18
5	A1-6	Heater Assembly	A2C-5	Control Unit	18
6	A1-5	Heater Assembly	A2C-1	Control Unit	18
7	A1-7	Heater Assembly	A2A-1	Control Unit	18
8	A1-8	Heater Assembly	A2A-1	Control Unit	18
9	A1-9	Heater Assembly	A2A-3	Control Unit	18
10	A1-10	Heater Assembly	A2A-5	Control Unit	18
11	A1-11	Heater Assembly	A2B-1	Control Unit	18
12	A1-12	Heater Assembly	A2B-2	Control Unit	18
13	E1-1	Fuel Pump	A2B-2	Control Unit	18
14	E1-2	Fuel Pump	A2A-4	Control Unit	18
15	HR-1	Glow Plug	GND	—	10
16	HR-2	Glow Plug	A2C-3	Control Unit	10
17	A2TEST-T	Control Unit	P6-17	Heater Switch	18
18	A2TEST-S	Control Unit	P6-24	Heater Switch	18
19	A2C-6	Control Unit	P6-16	Heater Switch	18
20	A2C-6	Control Unit	A2C-4	Control Unit	18
21	XF1-2	Fuse	A2C-4	Control Unit	10
22	XF1-1	Fuse	+24 VDC	—	10
23	E1-2	Fuel Pump	GND	—	18

Figure 39. Electrical Leads (Sheet 2 of 2).

(1) ITEM NO	(2) SMR CODE a. ARMY	b. AIR FORCE	(3) NSN	(4) CAGEC	(5) PART NUMBER	(6) DESCRIPTION AND USABLE ON CODE (UOC)	(7) QTY
						GROUP 14 WINTERIZATION KIT	
						FIG. 39 ELECTRICAL LEADS	
1	MFFFF	MOFFF		30554	98-2049-4	. LEAD, ELECTRICAL (NOT SHOWN) UOC: EMK, YNN	1
2	XBFZZ	XBFZZ	6145-00-851-8505	30554	88-20540-2	. . WIRE, ELECTRICAL MAKE FROM P/N M5086/2-20-9 (30554), 44 INCHES REQUIRED UOC: EMK, YNN	AR
3	XBFZZ	XBFZZ	5940-00-411-5913	30554	95-8105-2	. . TERMINAL, QUICK DISC UOC: EMK, YNN	2
4	XBFZZ	XBFZZ	5999-01-366-5119	30554	88-21941	. . PIN UOC: EMK, YNN	3
5	MFFFF	MOFFF		30554	98-2049-3	. LEAD, ELECTRICAL (NOT SHOWN) UOC: EMK, YNN	1
6	XBFZZ	XBFZZ	6145-00-851-8505	30554	88-20540-2	. . WIRE, ELECTRICAL MAKE FROM M5086/2-20-9 (30554), 44 INCHES REQUIRED UOC: EMK, YNN	AR
7	XBFZZ	XBFZZ	5940-00-813-0698	97403	13226E0107-3	. . TERMINAL, LUG UOC: EMK, YNN	1
8	MFFFF	MOFFF		30554	98-2050	. LEAD, ELECTRICAL (NOT SHOWN) UOC: EMK, YNN	1
9	XBFZZ	XBFZZ	6145-00-851-8505	30554	88-20540-2	. . WIRE, ELECTRICAL MAKE FROM M5086/2-20-9 (30554), 5 INCHES REQUIRED UOC: EMK, YNN	AR
10	XBFZZ	XBFZZ	5940-00-283-5280	97403	13226E0107-11	. . TERMINAL, LUG 1 UOC: EMK, YNN	1

END OF FIGURE

FIELD AND SUSTAINMENT MAINTENANCE

15 kW 50/60 AND 400 Hz SKID MOUNTED TACTICAL QUIET GENERATOR SETS

GROUP 15 BULK ITEMS:

(1)	(2) SMR CODE a. ARMY	b. AIR FORCE	(3)	(4)	(5)	(6)	(7)
ITEM NO	ARMY	AIR FORCE	NSN	CAGEC	PART NUMBER	DESCRIPTION AND USABLE ON CODE (UOC)	QTY

GROUP 15 BULK ITEMS

1 PAFZZ PAOZZ 6145-00-567-7412 16428 8407 . CABLE, POWER, ELECTRIC 1
2 PAFZZ PAOZZ 6145-01-213-5495 81349 M27500-16TE2T14 . CABLE, POWER, ELECTRIC 1
3 PAFZZ PAOZZ 6145-01-104-5076 81349 M27500-16TE1T14 . CABLE, POWER, ELECTRIC 1
4 PAFZZ PAOZZ 6145-01-129-9955 81349 M27500-20TE2T14 . CABLE, SPECIAL PURPOSE 1
5 PAFZZ PAOZZ 6145-01-128-2979 81349 M27500-20TE3T14 . CABLE, SPECIAL PURPOSE 1
6 PAFZZ PAOZZ 4010-01-365-6107 80244 RR-C-271TY2CL6 . CHAIN, WELDLESS .... 1 7 NOT USED
8 PAFZZ PAOZZ 4010-00-958-0633 81348RR-C-271TY2 CL7-16 . CHAIN, WELDLESS 1
9 PAFZZ PAOZZ 5325-00-074-3301 30554 88-20543-2 . GROMMET, NONMETALLIC .... 1 10 PAFZZ PAOZZ 5340-01-379-0688 03007 Z60960 . HINGE, ACCESS DOOR 72 INCHES
LONG 1
11 PAFZZ PAOZZ 5340-01-367-6703 03007 AA07060408 . HINGE, BUTT 1
12 PAFZZ PAOZZ 5340-01-367-6704 03007 AA06060406 . HINGE, BUTT 1
13 PAFZZ PAOZZ 4720-01-393-5581 96906 MS521302A101360 . HOSE, NONMETALLIC 1
14 PAFZZ PAOZZ 81349 100R1A-16 . HOSE, NONMETALLIC 1
15 PAFZZ PAOZZ 00624 303-5 . HOSE, NONMETALLIC 1
16 PAFZZ PAOZZ 4720-01-470-3929 30554 88-20579-3 . HOSE, NONMETALLIC 1
17 PAFZZ PAOZZ 81349 100R1AT-16 . HOSE, NONMETALLIC 1
18 PAFZZ PAOZZ 5970-00-822-2775 81349 M23053/5-110-9 . INSULATION SLEEVING 1
19 PAFZZ PAOZZ 5970-00-978-7677 81349 M23053/5-110-2 . INSULATION SLEEVING 1
20 PAFZZ PAOZZ 5970-00-954-1622 81349 M23053/5-105-0 . INSULATION SLEEVING 1
21 PAFZZ PAOZZ 5970-00-812-2969 81349 M23053/5-104-0 . INSULATION SLEEVING 1
22 PAFZZ PAOZZ 5970-00-082-3948 81349 M23053/5-108-9 . INSULATION SLEEVING 1
23 PAFZZ PAOZZ 5970-00-082-3942 81349 M23053/5-105-9 . INSULATION SLEEVING 1
24 PAFZZ PAOZZ 5970-01-367-0730 81349 M3190/03-17-0 . INSULATION, SLEEVING 1
25 PAFZZ PAOZZ 9390-01-408-1672 57137 62-B7-1/16 . NONMETALLIC CHANNEL 1
26 PAFZZ PAOZZ 4020-00-523-9641 72205 1-4INDIA . ROPE, FIBROUS 1
27 PAFZZ PAOZZ 9320-01-394-8045 81349MIL-R-6130 TYII/GRA . RUBBER STRIP 1
28 MFFZZ MOO 5330-01-369-8947 1X968 ZX-4134 . SEAL, NONMETALLIC 1
29 PAFZZ PAOZZ 5330-01-367-6329 30554 88-22705 . SEAL, NONMETALLIC SP 1
30 PAFZZ PAOZZ 5330-01-531-6822 30554 88-22708 . SEAL, NONMETALLIC SP .... 1 31 NOT USED
32 PAFZZ PAOZZ 5330-01-384-3057 30554 88-22712 . SEAL, PLAIN .... 1 33 PAFZZ PAOZZ 5330-01-384-1755 997396130TZS-S1SXPSA1S.75X1

.00 . SEAL, PLAIN 1

34 PAFZZ PAOZZ 5640-01-386-9618 28818 FF40JM02 . SOUND CONTROLLING B 24 X 54
INCHES 1
35 PAFZZ PAOZZ 6145-01-422-8602 30554 88-20444-2 . WIRE, ELECTRICAL 14 GA 1
36 PAFZZ PAOZZ 6145-01-422-4370 81343 SAE-J1127 . WIRE, ELECTRICAL 1
37 PAFZZ PAOZZ 6145-00-851-8505 81349 M5086/2-20-9 . WIRE, ELECTRICAL 1
38 PAFZZ PAOZZ 6145-00-578-6605 81349 M5086/2-16-9 . WIRE, ELECTRICAL 1
39 PAFZZ PAOZZ 6145-00-284-0657 81349 M5086/2-8-9 . WIRE, ELECTRICAL 1
40 PAFZZ PAOZZ 6145-00-578-7513 81349 M5086/2-10-9 . WIRE, ELECTRICAL 1
41 PAFZZ PAOZZ 6145-00-578-6596 81349 M5086/2-2-9 . WIRE, ELECTRICAL 1

(1)	(2) SMR CODE a.	b.	(3)	(4)	(5)	(6)	(7)
ITEM NO	ARMY	AIR FORCE	NSN	CAGEC	PART NUMBER	DESCRIPTION AND USABLE ON CODE (UOC)	QTY
42	PAFZZ	PAOZZ	4720-01-470-6230	30554	88-20579-4	HOSE, NONMETALLIC	1
43	PAFZZ	PAOZZ	4720-00-670-6037	30554	88-20579-6	HOSE, NONMETALLIC	1
44	PAFZZ	PAOZZ	4720-00-235-4131	30554	88-20579-7	HOSE, NONMETALLIC	1
45	PAFZZ	PAOZZ	5970-01-572-3826	18565	14-1214-0605-06CV	SHIELDING, CABLE	1
46	PAFZZ	PAOZZ	5975-01-572-0726	18565	14-1214-1602-00CV	SHIELDING, CABLE	1
47	PAFZZ	PAOZZ	5970-01-572-3824	18565	14-1214-0403-00CV	SHIELDING, CABLE	1

END OF FIGURE

FIELD AND SUSTAINMENT MAINTENANCE

15 kW 50/60 AND 400 Hz SKID MOUNTED TACTICAL QUIET GENERATOR SETS

SPECIAL TOOLS LIST

NOT APPLICABLE

END OF WORK PACKAGE

FIELD AND SUSTAINMENT MAINTENANCE

15 kW 50/60 AND 400 Hz SKID MOUNTED TACTICAL QUIET GENERATOR SETS

NATIONAL STOCK NUMBER INDEX

STOCK NUMBER	FIG.	ITEM
6625-00-003-0975	6 37 6625-00-	
004-8066	6 36	
5310-00-011-5093	1	25
	13	30
	24	2
	29	2
5310-00-011-6120	15	51
	25	15
5310-00-011-6121	26	19
	27	18
5306-00-017-6143	26	63
4720-00-021-3320	15	85
4730-00-024-3971	15	75
5305-00-036-6902	19	39
5305-00-036-6972	2	2
	7	18
	23	16
5305-00-038-3103	4	30
	7	44
5305-00-038-3145	7 33	
4730-00-041-2526	15	37
5310-00-044-6477	15	53
5306-00-050-0347	25	50
5340-00-050-2740	9 18	
5310-00-052-3632	4	31
	7	27
	9	1
	19	38
5340-00-053-8994	21	23
5340-00-057-3026	21	35
5305-00-059-4568	3 25	
5310-00-063-7360	2	1
	6	8
	7	17
	9	22
	23	26
4730-00-063-7919	15	31

STOCK NUMBER	FIG.	ITEM
5305-00-068-0500	25	2
	26	6
	26	31
5305-00-068-0501	26	26
5305-00-068-0502	25	9
	26	14
5305-00-068-0508	13	17
	15	96
5305-00-068-0509	13	33
	20	32
5305-00-068-0510	24	15
	27	21
	28	26
	29	4
5305-00-068-0511	24	4
5925-00-068-3931	7	9
5305-00-071-1319	6	9
5305-00-071-2066	27	17
5305-00-071-2075	27	16
	28	4
5305-00-071-2236	3 41	
5975-00-074-2072	7	82
	8	13
	13	48
	15	24
	15	59
	15	68
	16	9
	18	26
	21	12
	21	43
	21	78
5325-00-074-3301	BULK	9
5330-00-079-7840	7 80	
6625-00-081-5840	6 36	
5310-00-081-8087	7 24	
5970-00-082-3942	BULK	23
5970-00-082-3948	BULK	22

STOCK NUMBER	FIG.	ITEM
5310-00-087-7493	25	45
	26	57
4730-00-089-2515	11	1
5925-00-089-3031	7	8
6645-00-089-8842	6 33	
5975-00-111-3208	7	21
	8	12
	18	25
5940-00-113-0954	25	8
	26	13
5940-00-113-3140	19	66
5940-00-113-8184	8	3
	18	9
5940-00-113-8185	18	49
5940-00-113-8191	19	4
	19	8
	19	12
	19	15
5940-00-113-9820	25	7
	26	12
5940-00-113-9828	8	9
5940-00-113-9831	19	16
	19	20
	19	24
	19	28
5940-00-114-1300	25	37
5940-00-114-1305	18	6B
5940-00-114-1306	18	14
5940-00-114-1314	18	6A
5940-00-115-0763	18	13
5940-00-115-2677	9	6
	20	2
2590-00-141-9758	15	1
5940-00-143-4771	6	12
	18	12
	18	58
5940-00-143-4773	18	5
5940-00-143-4774	18	17
5940-00-143-4777	26	49

STOCK NUMBER	FIG.	ITEM
5940-00-143-4780	18	21
	18	57
5940-00-143-4794	18	10
5305-00-150-3408	7 23	
5961-00-154-7046	26	35
3110-00-155-6298	26	25
6240-00-155-7878	6	1
5325-00-174-9038	13	6
4730-00-177-6166	15	36
4730-00-187-0840	16	11
	15	17
5310-00-189-8467	20	39
5305-00-191-6226	19	31
4730-00-200-0531	15	42
5310-00-205-9951	20	30
5310-00-208-9255	21	55
5305-00-211-9344	2 58	
5305-00-218-4864	7	56
	7	88
5305-00-224-1092	7	63
	7	76
	19	61
5305-00-225-3843	13	55
	15	81
5306-00-225-8499	25	14
5305-00-225-8507	25	42
	26	54
5306-00-226-4827	15	52
	23	10
5315-00-227-6415	25	39
	26	51
5940-00-230-0515	8	87
	8	102
	8	133
	26	32
5935-00-233-3990	7 83	
4730-00-235-4066	37	4
4720-00-235-4131	BULK	44
4030-00-236-0843	13	10
5920-00-243-3787	7 93	

STOCK NUMBER	FIG.	ITEM
5305-00-253-5615	25	53
	26	66
5975-00-257-8055	4 19	
5305-00-269-2801	25	44
	26	56
5305-00-269-3233	1 24	
5305-00-269-3239	25	46
	26	58
5310-00-274-8710	2	6
	3	2
	4	14
	5	4
	7	90
	15	56
5310-00-274-8715	2	19
	3	29
	4	8
	10	5
	12	8
	13	26
	15	19
	20	33
	21	25
	25	10
	26	3
	26	27
	27	6
5305-00-282-3249	26	61
5940-00-283-5280	18	68
	39	10
5940-00-283-5281	8	4A
	18	48
	18	13A
6145-00-284-0657	18	22
	BULK	39
4730-00-287-3281	21	36
5325-00-290-1960	19	48
5305-00-292-4562	25	17
5961-00-295-5757	6	14
	7	69
	8	36
5340-00-297-0312	3	34
	4	11
	20	36
5935-00-315-9563	15	69
	18	41

STOCK NUMBER	FIG.	ITEM
5325-00-351-4543	19	47
5940-00-411-5913	39	3
5975-00-417-0543	7 19	
5945-00-458-3351	7 37	
5310-00-465-2719	20	9
5940-00-478-0037	7 95	
5961-00-478-7687	25	31
5935-00-482-7721	21	45
	21	80
5935-00-483-0259	15	61
	15	70
5306-00-484-5730	15	20
5940-00-504-4703	18	18
4020-00-523-9641	BULK	26
5310-00-533-2743	31	4
	33	14
	35	5
	37	5
5905-00-539-2479	7	1
5905-00-539-2573	6 20	
5920-00-539-6920	7	9
5305-00-543-4372	20	28
5940-00-549-6581	1	2
5940-00-549-6583	1	1
5310-00-550-1130	26	30
5340-00-550-5943	37	8
5905-00-556-5306	6 27	
5940-00-557-1629	8	8
	8	
5310-00-559-0070	26	38
5355-00-559-8943	6 10	
5935-00-565-9503	7 86	
6145-00-567-7412	BULK	1
5310-00-576-5752	25	24
6145-00-578-6596	BULK	41
6145-00-578-6605	8	18
	18	37
	BULK	38

STOCK NUMBER	FIG.	ITEM
6145-00-578-7513	18	24
	BULK	40
6145-00-578-7514	8	33
	18	23
5940-00-581-4325	1 28	
6210-00-583-9349	6	7
5940-00-606-4970	19	57
4730-00-613-6468	15	48
5940-00-615-6073	8	157
5930-00-615-6731	6 23	
4730-00-620-6904	15	46
5365-00-663-2125	7 85	
4720-00-670-6037	BULK	43
5940-00-682-2445	18	15
5330-00-684-7851	15	2
5310-00-696-5173	14	4
5340-00-724-7038	9	17
	19	42
5305-00-724-7265	24	8
5305-00-725-2317	27	3
4730-00-725-3664	16	13
5975-00-727-5153	15	67
5310-00-761-6882	21	24
	26	2
	26	29
5310-00-763-8920	26	64
5310-00-764-6609	24	10
4730-00-765-9103	15	49
5310-00-768-0318	25	51
4030-00-780-9350	9	5
4730-00-802-2560	15	5
5365-00-803-7304	25	22
5365-00-804-7654	26	47
5310-00-809-4058	14	5
5310-00-811-3494	13	35
4730-00-812-1333	15	22
5970-00-812-2969	8	37
	BULK	21

STOCK NUMBER	FIG.	ITEM
4730-00-812-7999	15	79
5940-00-813-0698	18	16
	39	7
5970-00-822-2775	BULK	18
5310-00-822-8525	2	37
	9	8
5935-00-823-5322	8 30 6210-00-	
5310-00-836-3520	7	62
	7	75
	8	20
	18	28
	19	60
5315-00-847-3531	26	48
6145-00-851-8505	8	15
	18	35
	39	2
	39	6
	39	9
	BULK	37
5935-00-852-9611	8	32
	18	91
5945-00-855-7478	21	3
6625-00-869-3141	6 34	
6625-00-869-3144	6 29	
5975-00-878-3791	3 40	
5920-00-892-9311	7 10	
4730-00-900-8663	22	8
5310-00-903-8595	3	7
	4	12
	9	13
4730-00-908-3194	22	2
4730-00-908-3195	13	4
	15	11
5310-00-913-8881	1 18	
5330-00-914-7651	7 79	
5310-00-929-1807	19	68
5320-00-932-1972	21	30
	32	2
5310-00-934-9757	26	37

STOCK NUMBER	FIG.	ITEM
5975-00-944-1499	8	14
	18	27
5970-00-954-1622	BULK	20
5940-00-954-3558	7 71	
5305-00-954-5629	9	2
5305-00-954-5638	9 33	
4010-00-958-0633	BULK	8
5940-00-958-1214	20	40
5970-00-978-7677	BULK	19
5305-00-983-7449	24	1
5310-00-984-3806	25	41
	26	53
5305-00-984-6189	25	28
5305-00-984-6196	26	36
5305-00-988-1727	26	42
5305-00-988-1728	26	4
5310-00-988-2652	21	60
5305-00-993-1849	25	29
4730-00-995-1559	15	41
5940-01-003-8579	19	56
	20	41
5935-01-012-1273	18	42
	18	42A
5310-01-012-3595	1	37
	2	8
	3	10
	4	1
	5	1
	7	92
	9	11
	13	7
	15	7
	17	1
	19	30
	20	5
	21	1
5961-01-013-0682	25	32
5935-01-014-4920	18	30
5935-01-014-7861	8 25	

STOCK NUMBER	FIG.	ITEM
5320-01-019-5694	2	64
	3	45
	4	27
	9	29
5935-01-020-4094	18	92
5999-01-039-8438	15	73
	18	40
	18	87
5935-01-044-8382	1 39 4730-01-	
5305-01-056-1501	4	17
	12	9
	13	27
	27	7
5905-01-063-9644	20	45
5961-01-067-9493	26	34
5340-01-071-8006	12	14
5340-01-074-8126	12	12
5980-01-076-8659	6	5
5342-01-078-9038	7	38
	7	40
5305-01-080-5721	26	21
5940-01-082-3321	18	59
	18	77
	20	11
	20	15
	20	19
	20	23
5340-01-086-2049	15	86
5999-01-092-2655	8	206
	15	60
	15	72
	21	44
	21	79
5310-01-099-9532	11	5
	13	61
	14	10
	21	71
	23	13
	27	10
	28	8
5935-01-101-7828	18	33

STOCK NUMBER	FIG.	ITEM
5935-01-102-7124	8 31	
2940-01-103-3267	12	18
2940-01-103-3268	12	17
5310-01-103-6042	2	20
	3	32
	10	7
	12	10
	13	28
	15	21
	19	50
	19	69
	20	34
	21	27
	27	8
6145-01-104-5076	BULK	3
5930-01-107-6474	21	8
	23	19
5340-01-107-7559	6 28	
5940-01-110-6423	7	26
	8	6
	8	62
	8	69
	8	77
	8	84
	8	92
	8	99
	8	107
	8	114
	8	122
	8	140
	8	148
	8	158
	8	167
	8	177
	8	199
	8	210
	18	8
5940-01-112-9746	18	11
	21	17
5935-01-114-5354	7 39	
4730-01-123-8618	15	29
4730-01-126-2173	15	87
5940-01-126-3973	21	18
6145-01-128-2979	BULK	5
6145-01-129-9955	BULK	4

STOCK NUMBER	FIG.	ITEM
4730-01-134-9827	15	6
5940-01-135-7081	8	41
	8	48
	8	55
	8	197
5940-01-136-0953	8 187 5940-01-	
5306-01-156-7663	1	38
	2	9
	3	6
	4	2
	5	2
	7	28
	7	89
	9	12
	13	8
	15	8
	17	2
	19	44
	20	6
	21	2
	31	1
	33	11
	35	1
5340-01-157-9475	4 18	
5325-01-161-2654	9 10	
5305-01-165-1254	15	55
5999-01-170-0558	18	31
	18	67
	18	76
	18	86
	38	1
5306-01-174-8738	21	76
	23	22
	28	16
5935-01-175-8419	7	54
	19	63
5306-01-179-1438	11	4
	14	11
	21	70
	23	12
	28	10
5310-01-183-5529	20	8
5305-01-187-5878	7	14
	7	74
	19	43

STOCK NUMBER	FIG.	ITEM
5310-01-190-4614	12	16
5310-01-195-6611	21	38
	21	67
5980-01-198-6311	6	6
5305-01-204-4683	7 77	
6145-01-213-5495	BULK	2
4720-01-215-0816	15	89
5310-01-215-7311	24	5
	25	3
5310-01-220-6513	21	41
5340-01-222-4225	19	49
5310-01-231-7459	27	13
5310-01-234-9415	3	31
	23	27
5310-01-234-9416	2	7
	3	3
	4	15
	5	5
	7	29
	15	57
	19	32
5315-01-236-0612	25	36
4730-01-236-1186	21	14
5905-01-236-4041	6 27	
5310-01-242-2679	24	6
5961-01-248-1712	25	25
5310-01-257-7590	1	19
	10	11
	13	45
	21	59
	27	2
	29	3
5940-01-259-2190	7	25
	8	11
	8	216
5310-01-266-4641	27	14
	28	22
5310-01-267-1685	15	4
5970-01-280-0362	7 41	
5305-01-283-8664	26	18
5310-01-289-7716	19	55
4720-01-293-4415	33	8

STOCK NUMBER	FIG.	ITEM
2920-01-298-6321	26	28
5325-01-301-7903	2	39
	9	16
5340-01-302-2960	25	27
	26	39
3110-01-304-8142	25	21
5310-01-307-7914	6 38	
4730-01-309-6370	15	95
5935-01-312-7038	8 22	
5999-01-320-7853	8 34	
5305-01-324-4896	25	23
2930-01-329-4857	25	40
	26	52
5305-01-334-8683	2	48
	19	59
5905-01-336-7533	7	6
4730-01-343-3192	15	28
5910-01-343-8827	19	67
2815-01-350-2207	27	12
6110-01-363-0493	7 55	
2910-01-364-9843	15	84
	16	8
5310-01-365-3190	27	15
	28	21
5310-01-365-4381	9	9
5310-01-365-4386	24	9
4010-01-365-6107	BULK	6
5305-01-365-6313	2	5
	3	1
	4	13
	5	3
5305-01-365-6314	20	44
5905-01-365-6585	7 65	
5310-01-365-8139	24	7
5305-01-365-9390	7 70	
5930-01-365-9613	7	7
5930-01-365-9614	21	29
	21	61
5925-01-365-9757	7 22 5945-01-	

STOCK NUMBER	FIG.	ITEM
5945-01-365-9954	7 42	
5930-01-366-0048	6 32	26 7
6625-01-366-0192	6 35 6625-01-	
5999-01-366-2621	7 8 18	59 16 38
5945-01-366-2725	7 35	
5945-01-366-2726	7 36	
5945-01-366-2727	7 48	
5945-01-366-2728	7 49	
5330-01-366-2836	15	65
5340-01-366-3360	24	12
5305-01-366-3501	24	11
5310-01-366-3539	10	9
5310-01-366-3540	21	51
5320-01-366-4394	3 4	23 3
5310-01-366-4412	19 20	54 29
5306-01-366-4527	1 21	
5999-01-366-5119	8 8 8 8 39	23 169 180 201 4
4720-01-366-6257	12	5
5330-01-366-6589	21	47
5315-01-366-6685	21	57
5905-01-366-7074	7 67	
5306-01-366-7075	3 4 21	33 7 26
4720-01-366-7172	15	40
5999-01-366-7952	18 18 18 18 18 18	32 46 56 66 75 85
5310-01-366-8134	7 16	

STOCK NUMBER	FIG.	ITEM
4730-01-366-9017	21	9
5935-01-366-9934	8 35	
5970-01-367-0730	BULK	24
5340-01-367-1503	10	1
4820-01-367-1836	21 23	11 21
5340-01-367-2287	21	39
5305-01-367-2314	8 18	21 29
5340-01-367-2328	21	54
5935-01-367-4422	8 28	
5330-01-367-6329	BULK	29
5340-01-367-6703	BULK	11
5340-01-367-6704	BULK	12
5935-01-367-7814	7 20	
6625-01-367-8436	7 60	
6110-01-367-8921	19	36
5340-01-367-8956	3	4
5940-01-367-9569	8 8 8 8 18	7 71 86
6695-01-367-9722	7 47	
6695-01-367-9723	21	19
2930-01-368-1071	13	46
6620-01-368-1531	6 32	
5905-01-368-2538	7 64	
5905-01-368-2539	7 66	
5905-01-368-2540	7 68	
5930-01-368-2891	6 22	
5930-01-368-2892	7	2
5930-01-368-2893	6 24	
5950-01-368-2915	19	33
5950-01-368-3006	19	35

STOCK NUMBER	FIG.	ITEM
5310-01-368-3048	10	10
	21	6
	21	22
	21	34
	23	23
	24	18
	28	11
	28	17
5310-01-368-3049	21	52
5360-01-368-4663	21	58
5930-01-368-5160	6 15	
4720-01-368-5430	13	15
5330-01-368-5932	10	13
5340-01-368-6048	2 35	
5340-01-368-6063	13	12
5940-01-368-6774	8 10	
6695-01-368-7114	7 31	
6110-01-368-7123	7 55	
2910-01-368-7644	7 50	
6625-01-368-7973	6 35	
5306-01-368-8041	2 55	
5945-01-369-0791	7 42	
2990-01-369-2079	10	3
5305-01-369-2166	11	13
	14	17
	14	26
	16	3
	28	28
5940-01-369-2268	1 33	
5940-01-369-2270	8	5
	8	72
	8	
	8	
	8	
	18	
5940-01-369-2271	19	
	19	
	19	
	19	
	19	
	19	

STOCK NUMBER	FIG.	ITEM
5940-01-369-2872	1	4
	1	8
	1	13
	1	27
	1	32
5940-01-369-2877	19	53
2910-01-369-5012	15	63
6685-01-369-6549	6 31	
5940-01-369-6948	18	19
	18	47
	20	12
	20	16
	20	20
	20	24
5330-01-369-7318	26	23
5330-01-369-8947	BULK	28
6695-01-369-9312	7 31	
2930-01-370-2868	13	63
5950-01-370-3328	19	35
4730-01-370-5426	13	16
2910-01-371-2689	21	46
2910-01-371-4356	7 50	
6110-01-372-2597	19	62
9905-01-372-7986	9 31	
5120-01-373-8976	20	4
4720-01-374-0783	13	2
5307-01-374-4451	2 38	
5330-01-374-4468	25	19
2990-01-374-9149	10	12
4720-01-375-1391	15	30
4720-01-375-1929	12	3
2910-01-376-2268	15	62
4730-01-376-4256	15	74
9905-01-377-5094	4 28	
5930-01-377-9113	21	32
4730-01-378-5224	13	22
5930-01-378-6921	21	10
	23	18
5340-01-379-0688	BULK	10

STOCK NUMBER	FIG.	ITEM
5305-01-379-5734	16	7
	28	14
6110-01-379-7187	7 55	
5305-01-380-3395	13	60
	21	5
	21	21
5365-01-381-3773	13	47
4820-01-381-5079	13	24
	14	20
5925-01-381-5199	7 22	
6625-01-381-7445	7 30	
6625-01-381-8195	7 34	
5340-01-382-2589	25	13
5950-01-382-3371	19	34
5999-01-382-8223	1 17	
5935-01-383-2612	8 29	
6115-01-383-3124	25	1
6110-01-383-4531	5	6
6150-01-383-6511	7	96
	8	1
6150-01-384-0013	7	96
	8	1
5940-01-384-0384	19	51
5910-01-384-1745	19	65
5330-01-384-1755	BULK	33
5330-01-384-3057	BULK	32
5365-01-384-9169	24	3
	25	47
	26	59
4720-01-385-1102	13	14
5930-01-385-1894	6 25	
6115-01-385-3104	26	22
5640-01-386-9618	BULK	34
2920-01-388-2776	4 20	
2910-01-388-6383	15	76
2990-01-389-3003	3	5
2990-01-390-4457	10	8
4720-01-392-0319	15	15
6680-01-392-8821	15	66

STOCK NUMBER	FIG.	ITEM
5340-01-392-8822	2 36	
5905-01-392-8825	21	15
5330-01-392-8826	21	50
4710-01-392-8846	12	4
5305-01-392-8847	21	68
4720-01-393-5581	BULK	13
5915-01-394-0942	20	48
4720-01-394-1931	15	39
9320-01-394-8045	BULK	27
5340-01-396-0454	2 59	
5920-01-396-1989	7 94	
5310-01-396-5836	3 26	
5915-01-396-9253	20	47
6110-01-397-2108	5	6
5305-01-406-1192	2 49	
5310-01-406-1672	2 45	
6150-01-406-9533	7 81	
4730-01-407-0649	15	93
6150-01-407-8102	18	1
	19	70
9390-01-408-1672	BULK	25
5920-01-414-6436	38	2
4730-01-417-5855	23	2
6140-01-418-6342	1	7
6145-01-422-4370	BULK	36
6145-01-422-8602	BULK	35
5305-01-423-8236	21	33

STOCK NUMBER	FIG.	ITEM
5940-01-425-2020	8	2
	8	2
	8	40
	8	47
	8	54
	8	64
	8	79
	8	94
	8	
	8	
	8	
	8	
	8	
	8	
	8	
	8	
	8	
	8	
	18	
5365-01-431-4540	2 46	
5365-01-431-4603	2 47	
6115-01-434-1432	26	1
6140-01-446-9498	1 22	
6140-01-457-4339	1 22	
5305-01-458-1624	13	29
5905-01-463-0058	20	22
5999-01-463-4017	8	24
	8	61
	8	131
	8	168
	8	179
	8	196
3030-01-463-9774	13	53
5305-01-464-6667	2 13	
5340-01-465-7761	15	9
5340-01-465-7765	15	10
6240-01-466-3528	6	3
5305-01-466-4406	13	34
5310-01-466-6312	2	18
	3	28
	4	16
	10	4
	12	7
	13	25
	15	18
	27	5
	28	1
5310-01-466-6687	13	44
	14	15
	29	1
5310-01-466-7247	19	52
	20	26
5310-01-466-7321	20	27
	21	69
5915-01-466-8148	20	48
5925-01-466-8153	7 22	
2930-01-466-8474	13	39
4140-01-466-8879	13	43
5305-01-466-9756	2 23	
5305-01-467-1561	9 23	
5342-01-467-4366	27	22
	28	20
9390-01-470-1205	13	58
5310-01-470-1286	15	50
	23	29
4730-01-470-1423	12	2
4730-01-470-1567	11	9
	13	13
	14	24
	23	17
4730-01-470-1578	10	2
	15	13
4730-01-470-1595	11	7
	12	1
	13	1
4730-01-470-1626	16	5
	23	3
4730-01-470-1701	15	14
4730-01-470-1982	22	6
4730-01-470-2409	23	1
5940-01-470-2470	18	36

STOCK NUMBER	FIG.	ITEM
5940-01-470-2768	19	41
5310-01-470-2776	28	23
5340-01-470-2973	7 45	
5940-01-470-3031	19	40
4720-01-470-3929	BULK	16
5961-01-470-4673	6 11	
5340-01-470-5184	37	1
4730-01-470-6199	33	10
4720-01-470-6230	BULK	42
5310-01-470-6529	28	24
4020-01-470-6597	20	1
4730-01-474-2274	15	35
4930-01-475-0388	15	44
5340-01-476-5376	21	40
6115-01-477-0566	30	6
5310-01-477-9621	23	28
5940-01-478-1154	18	39
4820-01-480-0846	15	43
6160-01-480-5918	1 20	
2915-01-482-8924	30	1
	31	5
2990-01-483-4051	30	3
	34	5
2910-01-483-4069	30	4
	33	4
5915-01-488-5004	20	47
5925-01-493-9106	7 22	
5310-01-494-2731	26	33
6150-01-494-3982	1 26	
6150-01-494-3983	1 12	
6150-01-494-3984	1	3
6150-01-494-3985	1 31	
6150-01-496-7717	30	2
	38	3
5930-01-499-3684	6 21	
5310-01-500-7620	21	20
4140-01-507-4174	13	59
4140-01-507-4176	13	54
5340-01-528-8654	4 25 5340-01-	
5305-01-531-4346	2	21
	3	30
	10	6
4730-01-531-4884	13	21
5310-01-531-4937	11	6
	14	9
	21	72
	23	14
	28	7
5310-01-531-4939	14	19
	14	27
	16	2
	21	66
5940-01-531-5519	8	63
	8	70
	8	78
	8	85
	8	93
	8	
	8	
	8	
	8	
	8	
	8	
	8	
	8	
	8	
	8	
	8	
	8	
	18	
	18	
	18	
	18	
	18	
5940-01-531-5660	7 57 5940-01-	
5310-01-531-6085	31	2
	33	12
	35	3
	37	7
5940-01-531-6448	8	4
	18	3
5310-01-531-6724	14	2

STOCK NUMBER	FIG.	ITEM
5330-01-531-6822	BULK	30
5340-01-531-7746	2 56	
5310-01-532-0291	31	3
	33	12
	35	4
	37	6
5310-01-532-0321	14	14
	23	24
	28	12
	28	18
5340-01-532-0643	2	3
4730-01-532-0694	15	3
5340-01-533-0084	2 26	
6130-01-533-2167	26	45
6115-01-533-2172	26	46
5340-01-535-0248	2 62	
4820-01-540-3024	15	47
5340-01-550-0921	14	7
6160-01-553-5219	1 23	
2910-01-553-6571	15	16
	16	12
1440-01-560-1702	21	42
4730-12-167-2104	34	4
2920-12-187-7756	34	1
4320-12-326-9400	34	3
5904-14-485-6282	34	2

END OF WORK PACKAGE

FIELD AND SUSTAINMENT MAINTENANCE

15 kW 50/60 AND 400 Hz SKID MOUNTED TACTICAL QUIET GENERATOR SETS

PART NUMBER INDEX

PART NUMBER	FIG.	ITEM
0102124405	34	1
01-21501-1	7	55
01-21501-2	7	55
01-21504-1 19 35		
01-21504-2 19 35		
01-21506-1	7	3
01-21506-2	7	3
01-21507-1	7	55
100R1A-16 BULK	14	
100R1AT-16	BULK	15
102064032050	34	4
11682345	1 39	
12-12 140339C	22	8
12 130109G	22	1
12258941	8 31	
13214E3291-3	4	31
13214E3789-2 21		30
	32	2
13218E0320-293	1 18	
13218E0320-49	31	4
	33	14
	35	5
	37	5
13226E0107-11	39	10
13226E0107-3 39	7	
1-4INDIA BULK	26	
150145	7 94	
1901-0759	6	14
	7	69
	8	36
19450-2	7	22
1N3663	25	31
1N3663R	25	32
2-130109E	21	36

PART NUMBER	FIG.	ITEM
2-2 140239C	16	13
251667010001	34	2
251671250100	34	3
29100	7 22	
312KDD	26	25
4-130109E-C	11	1
4-2 070102C	15	17
4-2 140137C	15	5
4-4 070202C	16	6
4-4-130238	15	48
4-4-4-140424B	33	10
4-4-4 140424C	15	36
6130TZS-S1SXPSA1S.75X1.	BULK	33
62-B7-1/16 BULK	25	
6308-2RSJEM	25	21
69-539-2 15 22		
69-561-1	7	62
	7	75
	8	20
	18	28
	19	60
69-561-2	2	1
	6	8
	7	17
	9	22
	23	26
69-561-3	7	27
	9	1
	19	38
69-561-5 14	4	

PART NUMBER	FIG.	ITEM
69-561-6	1	37
	2	8
	3	10
	4	1
	5	1
	7	92
	9	11
	13	7
	15	7
	17	1
	19	30
	20	5
	21	1
69-570-3	4	19
69-574	6 34	
69-583	4 18	
69-593	6	2
69-594	6	1
69-597	6 36	
69-598	6 36	
69-599	6 29	
69-662-11	8	21
	18	29
69-662-17	2 58	
69-662-20	2	2
	7	18
	23	16
69-662-22	7	56
	7	88
69-662-35	4	30
	7	44
69-662-36	19	39
69-662-5	7	63
	7	76
	19	61
69-662-64	19	31
69-662-65	7	74
	19	43
69-662-7	7	77
69-668	15	85
69-668-2	15	88
69-692-1	20	40

PART NUMBER	FIG.	ITEM
6TMF/TYPEI	1 22 70-1580	
702806-01	25	48
	26	60
703320-02	25	16
703513-01	26	20
70-4012	6 37	
706336-03	25	55
707329-01	25	38
	26 50	
707806-0A	25	34
707807-02	26	45
716327-0A	25	40
	26	52
718337-01	26	17
718514-01	26	7
718515-01	26	9
718516-01	26	8
718517-01	26	10
718810	25	26
720344-03	25	6
720345-03	25	5
720346-0C	25	4
720982-0A	26	5
72-2062-1 19 55		
72-2098-2 19 49		
72-2173-1 15 49		
72-2213	19	57
72-2236	19	56
	20	41
72-5304	22	6
73-0507	6 33	
75-0809	15	87
752818-0A	25	11
	26 15	
777056-0A	26	28
777068-0A	26	24
777070-0A	25	20

PART NUMBER	FIG.	ITEM
777079-0A	25	25
778715-0A	25	33
783621-0A	25	49
783622-0A	26	62
789293-0A	26	22
789295-0A	25	18
791149-0A	25	35
791150-0A	26	46
801016-06	25	23
834822-01	25	12
	26	16
8407	BULK	1
8-4 430160C	23	2
846833-01	25	13
865873-01	26	23
865876-01	25	19
88-20013	2 57	
88-20014	2 56	
88-20018-1 16	9	
88-20024-5 15 86		
88-20029-4 15 31		
88-20033-10A	31	2
	33	12
	35	3
	37	7
88-20033-21A	11	15
	28 3	
88-20033-22A	14	19
	14	27
	16	2
	21	66
88-20033-24A 11		6
	14	9
	21	72
	23	14
	28	7
88-20033-26B 20 30		
88-20033-31A 14		14
	23	24
	28	12
	28	18

PART NUMBER	FIG.	ITEM
88-20033-34A	28	25
88-20033-40B	20	9
88-20033-41A	28	22
88-20037WDG21	33	9
88-20043	15	43
88-20049	15	44
88-20050	2	39
	9	16
88-20051	9 10	
88-20052	2	37
	9	8
88-20063-05	3 46	
88-20063-06	3 46	
88-20064-5 25 54		
88-20064-6 26 67		
88-20073	9 32	
88-20074	9 30	
88-20075	3 47	
88-20102	4 28	
88-20110	2 67	
88-20123	2 24	
88-20126	2 68	
88-20147-1	7	9
88-20176	6 24	
88-20187	13	58
88-20188	1 17	
88-20197	9 31	
88-20210	3	23
	4	3
88-20218	4 20 88-	
88-20247-2 27		15
	28	21
88-20247-3 24	7	
88-20251	7 20	
88-20252	2 59	
88-20258	7 45	
88-20260-20	2 55	

PART NUMBER	FIG.	ITEM
88-20260-21	1	38
	2	9
	3	6
	4	2
	5	2
	7	28
	7	89
	9	12
	13	8
	15	8
	17	2
	19	44
	20	6
	21	2
	31	1
	33	11
	35	1
88-20260-23	2	5
	3	1
	4	13
	5	3
88-20260-25	20	44
88-20260-30	11	13
	14	17
	14	26
	16	3
	28	28
88-20260-31	15	20
	2	21
	3	30
	10	6
88-20260-32	4	17
	12	9
	13	27
	27	7
88-20260-33	3	33
	4	7
	21	26
88-20260-35	16	7
	28	14
88-20260-51	14	3
88-20262	7 64	
88-20263	7 65	
88-20265	7 66	
88-20266	7 67	
88-20267	7 68	

PART NUMBER	FIG.	ITEM
88-20272	7 71	
88-20274-1	8	2
	8	2
	8	40
	8	47
	8	54
	8	64
	8	79
	8	
	8	
	8	
	8	
	8	
	8	
	8	
	8	
	8	
	18	
88-20274-11	7	
	8	
	8	
88-20274-3	8	
	18	
88-20274-4	8	
	8	
	8	
	8	
	18	
88-20274-5	7	
	8	
	8	
	8	
	8	
	8	
	8	
	8	
	8	
	8	
	8	
	8	
	8	
	8	
	8	
	8	
	18	
88-20274-6	8	
	8	
	8	
	8	
	18	

PART NUMBER	FIG.	ITEM
88-20275-1 18		11
	21	17
88-20275-2 21 18		
88-20275-3 18		59
	18	77
	20	11
	20	15
	20	19
	20	23
88-20275-4 18 78		
88-20276	2 11	
88-20277	2 10	
88-20278	2	3
88-20286	15	65
88-20299	7 70	
88-20305-1 20 10		
88-20305-1-1 20 13		
88-20305-2 20 14		
88-20305-2-1 20 17		
88-20305-3 20 18		
88-20305-3-1 20 21		
88-20305-5 20 22		
88-20305-5-1 20 25		
88-20444-2 BULK	35	
88-20461	2 50	
88-20468	13	63
88-20468-1 13 64		
88-20468-2 13 65		
88-20468-3 13 66		
88-20470	18	42
	18	42A
88-20471	21	45
	21	80
88-20472	18	92
88-20474	15	69
	18	41
88-20475	15	61
	15	70

PART NUMBER	FIG.	ITEM
88-20476	8	206
	15	60
	15	72
	21	44
	21	79
88-20477	15	73
	18	40
	18	87
88-20492	15	66
88-20540-2 39		2
	39	6
	39	9
88-20543-2 BULK	9	
88-20551-1 15 89		
88-20554-1 21 75		
88-20556-38	11	14
	14	18
	14	28
	16	1
	28	2
88-20556-4 31		3
	33	13
	35	4
	37	6
88-20556-6 23 28 88-20556-7 14	2	
88-20561-1 16		5
	23	3
88-20561-2 23	1	
88-20561-3 11		9
	13	13
	14	24
	23	17
88-20561-4 11		7
	12	1
	13	1
88-20561-5 12	2	
88-20561-6 10		2
	15	13
88-20561-7 15 14		
88-20564-2 14	5	
88-20564-3 23 27		

PART NUMBER	FIG.	ITEM
88-20563-4 28 24		
88-20563-5 28 23		
88-20574-2	8	209
88-20575	7 10	
88-20579-3 BULK	16	
88-20579-4 BULK	42	
88-20579-6 BULK	43	
88-20579-7 BULK	44	
88-20592-1	7	1
88-20596-1	8	63
	8	70
	8	78
	8	85
	8	93
	8	100
	8	108
	8	115
	8	123
	8	130
	8	141
	8	149
	8	156
	8	166
	8	176
	8	198
	8	211
	18	45
	18	55
	18	65
	18	74
	18	84
88-20596-2	8	41
	8	48
	8	55
	8	197
88-20596-3	8	187
88-20879-7 23	7	
88-21005	25	1
88-21006	26	1
88-21015	19	34
88-21043	15	76
88-21061	21	19
88-21065	12	13
88-21066	19	36

PART NUMBER	FIG.	ITEM
88-21069-12	24	12
88-21071-3 27		22
	28	20
88-21072	6 35	
88-21073	6 35	
88-21078	6 22	
88-21080	7	7
88-21081	21	29
	21	61
88-21082	19	37
88-21084	8 28	
88-21085	7 37	
88-21098	2 35	
88-21099	2 15	
88-21101	15	84
	16	8
88-21103	6 30	
88-21104	6 32	
88-21105	6 31	
88-21114-2 15 37		
88-21121-1 15 62		
88-21124	6	4
88-21126	21	10
	23	18
88-21127	12	6
88-21129	12	12
88-21131	19	33
88-21133	7 47	
88-21134	7 31	
88-21135	7 31	
88-21138	7 35	
88-21140	7 49	
88-21141	7 36	
88-21142	7 30	
88-21144	7 60	
88-21145	7 48	
88-21146	20	4
88-21155	7	8

PART NUMBER	FIG.	ITEM
88-21162-3 15 63		
88-21165	7 34	
88-21173	15	42
88-21177-1 12	3	
88-21179	20	45
88-21181-10	18	39
88-21181-12	8 17	
88-21181-2 18 36		
88-21181-22	8 19	
88-21182-10	19	40
88-21182-12	7 58	
88-21182-2 19 41		
88-21182-22	7 57	
88-21183-2	7	59
	8	16
	18	38
88-21580	19	1
88-21582	19	1
88-21585	2 31	
88-21590	3 43	
88-21593	7 15	
88-21596	3 19	
88-21598	3	8
88-21600	3 13	
88-21601	3 20	
88-21603	2 71	
88-21604-104	15	33
88-21604-110	22	3
88-21604-146	15	32
88-21604-151	15	34
88-21604-228	13	3
88-21604-272	13	11
88-21604-295	2	4
88-21604-317	3 22	
	13 50	
88-21604-339	15	77
88-21604-343	15	38

PART NUMBER	FIG.	ITEM
88-21604-356	2 16	
88-21604-364	13	37
88-21604-396	15	94
88-21605-104	15	33
88-21605-110	22	3
88-21605-146	15	32
88-21605-151	15	34
88-21605-228	13	3
88-21605-272	13	11
88-21605-295	2	4
88-21605-317	3 22	
	13 50	
88-21605-339	15	77
88-21605-343	15	38
88-21605-356	2 16	
88-21605-364	13	37
88-21605-396	15	94
88-21633	21	11
	23	21
88-21635	2 65	
88-21649-17	26	41
88-21649-19	26	40
88-21666	20	38
88-21669	19	58
88-21674-1	7	16
88-21674-2	3	7
	4	12
	9	13
88-21674-3	3	34
	4	11
	20	36
88-21676-1	7	6
88-21683	1 23	
88-21684	1 20	
88-21685	1 21	
88-21694	3 36	
88-21695	13	43
88-21700	13	31

PART NUMBER	FIG.	ITEM
88-21712	10	8
88-21723	4 29	
88-21729	8 35	
88-21730	8 34	
88-21731	29	5
88-21732	2 40	
88-21733	2 41	
88-21743	12	11
88-21755-3 22	4	
88-21755-4 16 11		
88-21767	20	35
88-21770	4 25	
88-21771	2 33	
88-21775	20	46
88-21776	21	31
88-21783	25	43
	26	55
88-21795-8 25 30		
88-21798	26	68
88-21804	26	11
88-21813	2 63	
88-21814	2 61	
88-21815	2 62	
88-21820	2 32	
88-21821-01	3 48	
88-21821-02	3 48	
88-21849	13	9
88-21850	15	64
88-21854	15	54
88-21860	7 50	
88-21861	7 50	
88-21868	9 14	
88-21869	9 15	
88-21870	2 12	
88-21871	9 26	
88-21873	9 25	
88-21874	9 24	

PART NUMBER	FIG.	ITEM
88-21875	2 17	
88-21876	9 21	
88-21877	2 14	
88-21883	15	35
88-21884	15	6
88-21887	21	14
88-21889	3 42	
88-21890	2 36	
88-21891	22	5
88-21892	15	9
88-21893	15	10
88-21896	7 42	
88-21897-1	8	29
88-21897-2	7	43
88-21898	6	6
88-21899	6	5
88-21913	3 35	
88-21925	4 10	
88-21928	3 18	
88-21930-1 19		54
	20	29
88-21930-2 20	8	
88-21930-3	3	26
88-21930-4	2	45
88-21932	19	45
88-21933	20	37
88-21939	8 25	
88-21940	18	34
88-21941	8	23
	8	169
	8	201
	39	4
88-21942	18	31
	18	67
	18	76
	18	86
	38	1

PART NUMBER	FIG.	ITEM
88-21943	8	24
	8	61
	8	131
	8	179
	8	196
88-21944	18	32
	18	46
	18	56
	18	75
	18	85
88-21947	18	30
88-21952	3 39	
88-21959	10	3
88-21962	3 12	
88-21964	4 21	
88-21971	7 73	
88-21975	10	1
88-21976	7 15	
88-21977	4 22	
88-21980	9 19	
88-21981	8 27	
88-21982	18	33
88-21988	3	9
88-21995	7 87	
88-21998	2 60	
88-22007	29	6
88-22032	3 44	
88-22037	4 26	
88-22039	19	64
88-22040	4 24	
88-22042	9 20	
88-22051	3 37	
88-22060	19	46
88-22063	2 43	
88-22068	15	15
88-22070	2 30	
88-22071	8 10	
88-22072	4	4
88-22074	4 23	

PART NUMBER	FIG.	ITEM
88-22077	2 22	
88-22078	2 65	
88-22088	2 28	
88-22089	2 29	
88-22096-2 15 47		
88-22098-1	8	38
88-22098-1-1	8	39
88-22098-2	8	45
88-22098-2-1	8	46
88-22098-3	8	52
88-22098-3-1	8	53
88-22099-1	8	59
88-22099-1-1	8	60
88-22099-2	8	67
88-22099-2-1	8	68
88-22099-3	8	75
88-22099-3-1	8	76
88-22099-5	8	90
88-22099-5-1	8	91
88-22099-6	8	97
88-22099-6-1	8	98
88-22099-7	8	82
88-22099-7-1	8	83
88-22100	8 105	
88-22100-1	8	106
88-22101-1	8	112
88-22101-10	18	43
88-22101-10-1 18 44		
88-22101-11	18	53
88-22101-1-1	8	113
88-22101-11-1 18 54		
88-22101-2	8	120
88-22101-2-1	8	121
88-22101-3	8	128
88-22101-3-1	8	129
88-221014	8 138	
88-22101-4-1	8	139

PART NUMBER	FIG.	ITEM
88-22101-5	8	146
88-22101-5-1	8	147
88-22102-1	8	154
88-22102-1-1	8	155
88-22102-1-12	8	160
88-22102-2	8	164
88-22102-2-1	8	165
88-22102-3	8	174
88-22102-3-1	8	175
88-22103-7 18 63		
88-22103-7-1 18 64		
88-22103-8 18 72		
88-22103-8-1 18 73		
88-22103-9 18 82		
88-22103-9-1 18 83		
88-22104	8	184
88-22104-1	8	185
88-22106	7 61	
88-22106-11	7 72	
88-22111	15	12
88-22116	3 11	
88-22119-11	1 33	
88-22119-12	18	19
	18	47
	20	12
	20	16
	20	20
	20	24
88-22119-14	1	4
	1	8
	1	13
	1	27
	1	32
88-22119-22	1 28	
88-22119-3 19		3
	19	7
	19	19
	19	23
	19	27
88-22120	9	3
88-22120-4	9	4
88-22122	19	47
88-22123	1	3
88-22123-1	1	6
88-22123-3	1	5
88-22126-1 19	2	
88-22126-1-1 19	5	
88-22126-2 19	6	
88-22126-2-1 19	9	
88-22126-3 19 10		
88-22126-3-1 19 13		
88-22126-4 19 14		
88-22126-4-1 19 17		
88-22126-5 19 18		
88-22126-5-1 19 21		
88-22126-6 19 22		
88-22126-6-1 19 25		
88-22126-7 19 26		
88-22126-7-1 19 29		
88-22135	27	19
88-22136	20	7
88-22137	5	6
88-22137-105	6 16	
	7 11	
88-22138	5	6
88-22141	6	26
	32	7
88-22146	20	43
88-22163-1 20 42		
88-22164	18	1
	19 70	
88-22167	13	46
88-22170	13	2
88-22171	13	14
88-22172	13	15
88-22174	27	20
88-22179	1	7

PART NUMBER	FIG.	ITEM
88-22179-1	1	11
88-22179-3	1	9
88-22179-4	1	10
88-22182	10	13
88-22183	10	12
88-22201	21	20
88-22202	21	3
88-22209	7	81
88-22209-7	7	84
88-22210	2	69
88-22211	2	66
88-22300	12	5
88-22301	12	4
88-22302-12	13	5
88-22302-4	13	18
88-22302-6	13	19
88-22303	7	96
	8	1
88-22307	21	9
88-22309	1	26
88-22309-1	1	30
88-22309-3	1	29
88-22310	1	12
88-22310-1	1	16
88-22310-3	1	14
88-22310-4	1	15
88-22311	1	31
88-22311-1	1	36
88-22311-4	1	34
88-22311-5	1	35
88-22312	8	22
88-22313	8	26
88-22315	27	12
88-22317	21	32
88-22319	21	7
88-22320	21	28
88-22321	27	4

PART NUMBER	FIG.	ITEM
88-22322	21	15
	21	64
88-22323	13	21
88-22324	15	74
88-22325	15	40
88-22326-1	15	39
88-22331-1	11	5
	13	61
	14	10
	21	71
	23	13
	27	10
	28	8
88-22331-2	10	10
	21	6
	21	22
	21	34
	23	23
	24	18
	28	11
	28	17
88-22331-3	21	38
	21	67
88-22331-4	21	52
88-22336-1	20	39
88-22338-1	15	30
88-22339	24	3
	25	47
	26	59
88-22416	21	50
88-22418-2	6	11
88-22418-2-2	6	13
88-22424	13	32
88-22425	27	11
88-22428	21	46
88-22429	3	5
88-22430	3	4
88-22436	19	53
88-22437	19	51
88-22459	20	31
88-22463	7	42

PART NUMBER	FIG.	ITEM
88-22469	20	1
88-22469-2 20	3	
88-22472	2 26	
88-22474	2 52	
88-22475	2 54	
88-22476	2 53	
88-22478	2 44	
88-22479	2 51	
88-22480	2 70	
88-22481	21	13
88-22482	2 47	
88-22483	2 46	
88-22485	15	80
88-22486	15	78
88-22489	13	59
88-22490	13	54
88-22491	13	57
88-22492	13	56
88-22493	13	62
88-22500	2 27	
88-22501	2 25	
88-22508	24	9
88-22509	19	62
88-22510	3 17	
88-22512	3 21	
88-22513	3 14	
88-22514	3 38	
88-22524	24	13
88-22525	24	14
88-22528	3 15	
88-22530	24	11
88-22533	13	12
88-22543	13	67
88-22545	21	42
88-22547	15	58
88-22550	21	16
88-22554-2 13 16		

PART NUMBER	FIG.	ITEM
88-22554-3 13 22		
88-22555	6 28	
88-22580	7	2
88-22594	4	5
88-22595	3 24	
88-22617	21	49
88-22618	21	47
88-22619	21	54
88-22620	21	57
88-22621	21	56
88-22623	21	39
88-22624	21	58
88-22626	2 38	
88-22629-002	8 43	
88-22629-004	8 42	
88-22629-005	8 44	
88-22629-006	8 49	
88-22629-007	8 50	
88-22629-008	8 51	
88-22629-009	8 56	
88-22629-010	8 57	
88-22629-011	8 58	
88-22629-012	8 65	
88-22629-013	8 66	
88-22629-014	8 73	
88-22629-015	8 74	
88-22629-016	8 80	
88-22629-017	8 81	
88-22629-019	8 88	
88-22629-020	8 95	
88-22629-021	7 46	
	8 96	
88-22629-022	8 103	
88-22629-023	8 104	
88-22629-024	8 89	
88-22629-025	8 110	
88-22629-026	8 111	

PART NUMBER	FIG.	ITEM
88-22629-027	8	117
88-22629-028	8	118
88-22629-029	8	119
88-22629-030	8	125
88-22629-031	8	126
88-22629-032	8	127
88-22629-033	8	135
88-22629-034	8	136
88-22629-035	8	137
88-22629-036	8	143
88-22629-037	8	144
88-22629-038	8	145
88-22629-039	8	151
88-22629-040	7 13	
	8 152	
	8 214	
88-22629-041	7 12	
	8 153	
	8 215	
88-22629-043	18	50
88-22629-044	18	51
88-22629-046	18	60
88-22629-050	18	52
88-22629-051	18	61
88-22629-052	18	62
88-22629-053	8	161
88-22629-054	8	162
88-22629-055	8	163
88-22629-056	8	171
88-22629-057	8	172
88-22629-058	8	173
88-22629-059	8	181
88-22629-060	8	182
88-22629-061	8	183
88-22629-063	18	69
88-22629-064	18	70
88-22629-066	18	79

PART NUMBER	FIG.	ITEM
88-22629-067	18	80
88-22629-069	18	88
88-22629-070	18	89
88-22629-074	18	71
88-22629-075	18	81
88-22629-076	18	90
88-22629-077	8 189	
88-22629-078	6 17	
	8 190	
88-22629-079	6 18	
	8 191	
88-22629-080	6 19	
	8 192	
88-22629-082	8 203	
88-22629-084	8 205	
88-22629-085	8 202	
88-22629-086	8 204	
88-22629-087	8 213	
88-22630	8 193	
88-22630-1	8	194
88-22630-2	8	195
88-22643	15	97
88-22654	15	98
88-22661	4	6
88-22664	7	96
	8	1
88-22666	2 42	
88-22667	3 27	
88-22668	13	39
88-22672	7 39	
88-22701	17	3
88-22702	3 16	
88-22703	13	47
88-22704	17	4
88-22705	BULK	29
88-22706	21	8
	23	19

PART NUMBER	FIG.	ITEM
88-22708	BULK	30
88-22712	BULK	32
88-22722	13	52
88-22723	13	49
88-22724	13	51
88-22725	13	40
88-22726	13	41
88-22727	13	36
88-22728	13	42
88-22729	13	38
88-22737	9 32	
88-22738	2 34	
88-22740	13	53
88-22748	15	92
88-22751	13	24
	14	20
88-22752-2 13 23		
88-22752-3 13 20		
88-22757-2 19 67		
88-22758	19	65
88-22761	20	47
88-22762	20	48
88-22765	15	93
88-22767	7 41	
88-22768	7	38
	7	40
88-22772	9 27	
88-22773-1	6	20
88-22773-2	6	27
88-22773-3	6	27
88-22774	9 28	
88-22784	21	40
88-22786	19	52
	20	26
88-22790-1	2	18
	3	28
	4	16
	10	4
	12	7
	13	25
	15	18
	27	5
	28	1
88-22790-2 15		50
	23	29
88-22790-3 13		44
	14	15
	29	1
88-22791-1	2	23
88-22792-2 20		27
	21	69
88-22793-4	2	13
88-22795	15	3
88-22800	1 22	
88-22805	20	47
88-22807	20	48
88-22825-2	7	22
88-22826-2	7	22
88-24002-1 16 10		
95-8105-2 39	3	
95-8125-4 21 55		
96-23615	37	1
96-23737	6 21	
97-24002	23	8
97-24010	28	5
97-24011	22	7
97-24012	28	27
97-24013	28	15
97-24014	28	19
97-24015	23	25
97-24016	23	20
97-24017	21	73
97-24018	21	81

PART NUMBER	FIG.	ITEM
97-24019	14	8
97-24020	11	16
97-24021	11	2
97-24022	11	8
97-24023	28	6
97-24024	23	11
97-24025	23	9
97-24026	23	15
97-24028	11	11
97-24029	11	10
97-24030	11	12
97-24031	23	5
97-24032	14	23
97-24033	14	29
97-24034	14	1
97-24035	14	25
97-24036	14	12
97-24037	28	30
97-24038	28	13
97-24039	28	29
97-24040	14	16
97-24041	14	22
97-24042	12	11
97-24044	16	15
97-24045	21	62
97-24048	18	1A
	19	70
97-24049	7 96	
	8 1	
97-24050-01	3 46	
97-24050-02	3 46	
97-24051	2 69	
97-24052	2 66	
97-24053	2 65	
97-24060	7 96	
	8 1	
97-24061	21	63

PART NUMBER	FIG.	ITEM
97-24071	16	4
	16 16	
97-24080	5	6
97-24081	5	6
97-24082	19	1
97-24110	7	9
97-24111-9 11	3	
97-24113	23	6
97-24114	21	77
97-24116-4 24 20		
97-24116-5 24		17
	28	31
97-24116-6 28 31		
97-24118	21	65
97-24120	7 15	
97-24121	7 15	
97-24122	24	19
97-24124	2 72	
97-24128	2 76	
97-24129	2 77	
97-24130	2 73	
97-24131	2 75	
97-24132	2 78	
97-24133	8 208	
97-24134	7 50	
97-24137	2 74	
97-24138	2 31	
97-24146	14	21
97-24604-104	15	33
97-24604-110	22	3
97-24604-112	23	7
97-24604-146	15	32
97-24604-151	15	34
97-24604-228	13	3
97-24604-272	13	11
97-24604-282	23	4
97-24604-295	2	4

PART NUMBER	FIG.	ITEM
97-24604-317	3 22	
97-24604-339	15	77
97-24604-343	15	38
97-24604-356	2 16	
97-24604-396	15	94
97-24605-104	15	33
97-24605-110	22	3
97-24605-112	23	7
97-24605-146	15	32
97-24605-151	15	34
97-24605-228	13	3
97-24605-272	13	11
97-24605-282	23	4
97-24605-295	2	4
97-24605-317	3 22	
97-24605-339	15	77
97-24605-343	15	38
97-24605-356	2 16	
97-24605-396	15	94
98-19744-03	16	14
98-2000	30 34	3 5
98-2001	32	3
98-2002	35	2
98-2003	32	10
98-2008-4 37	3	
98-2013	30 33	4 4
98-2014	30 31	1 5
98-2015	30	6
98-2017-3 37	8	2 3
98-2019-2 37	2	
98-2021	32	9
98-2024-5 37	4	
98-2025-3 33	6	

PART NUMBER	FIG.	ITEM
98-2041	32	1
98-2043	32	5 5
8		4
98-2049-3 32	39	1
98-2049-4 32		6 8
98-2053-3 30	5 A50452-1	
6	3	
A52425-1	1	2
A-9696	25 26 65	52
AA06060406	BULK	12
AA07060408	BULK	11
AA-52425-2	1	1
AA55569/02-009 38	2	
AA55804-3B 9FT	3	40
AN315-4 26 33		
AN565C416-12	25	17
AN565C416H6	26	21
AN565C816H16 26 61		17
B1821BH025C075N	13 15	96 55
B1821BH025C100N	13 15	81 33
B1821BH025C125N	13 20	32 52
B1821BH031C100N	15	10
B1821BH038C062N	13	29
B1821BH038C075N	20	28
B1821BH038C100N	24 27 28 29	15 21 26 4
B1821BH038C125N	24	4
B1821BH038C150N	27	3
B1821BH038F063N	1	24

PART NUMBER	FIG.	ITEM
B1821BH038F138N	25	46
	26	58
B1821BH050C100N	27	17
B1821BH050C300N	27	16
	28	4
B1821BH063C475N	24	8
B18231A03014	28	9
B18231A10020N	21	76
	23	22
	28	16
B18231A10025N	11	4
	14	11
	21	70
	23	12
	28	10
B18231B10016N	21	33
B18231B10025N	13	60
	21	5
	21	21
B18231B5016N 21 53		
B18231B6020N 21 48		
B18231B8025N 27	9	
B18241A060	21	41
B18241B050	21	51
B18241B100	10	9
B1831BH060014	21	37
B1831BH060016	21	68
B-718817-01	26	44
B-718930-01	26	43
DIN73378-4x1.25-PA12-HL-nf	33	7
DIN73378-6x2-PA12-HL-sw	33	3
F03A250V10AS	7	93
FF40JM02 BULK	34	
FFS92TYPE3STYLE2C	6	9
J1508-10 TYPE D 33	1	
J1508-13 TYPE D 33	5	
J30R7-TYPE 1-3/16ID 33	8	
J30R7-TYPE 1-5/32ID 33	2	
JANTX1N1190	26	35
JANTX1N1190R	26	34

PART NUMBER	FIG.	ITEM
M23053/5-104-0	8	37
	BULK	21
M23053/5-105-0 BULK	20	
M23053/5-105-9 BULK	23	
M23053/5-108-9 BULK	22	
M23053/5-110-2 BULK	19	
M23053/5-110-9 BULK	18	
M24243/1A402	2 64	
M27500-16TE1T14	BULK	3
M27500-16TE2T14	BULK	2
M27500-20TE2T14	BULK	4
M27500-20TE3T14	BULK	5
M3190/03-17-0 BULK	24	
M5086/2-10-9 BULK	40	
M5086/2-12-9	8	33
	18	23
M5086/2-16-9	8	18
	18	37
	BULK	38
M5086/2-20-9 8 18		15
	BULK	35
		37
M5086/2-2-9 BULK	41	
M5086/2-8-9 18		22
	BULK	39
MIL-R-6130TYII/GRA BULK	27	
MS16624-1156	25	22
MS16624-1300	26	47
MS16998-75	24	1
MS20066-352	25	36
MS20066-354	25	39
	26	51
MS20066-356	26	48
MS20659-105	25	37
MS20659-125	19	66
MS20659-128	25	7
	26	12
MS20659-129	18	6A

PART NUMBER	FIG.	ITEM
MS20659-144	9	6
	20	2
MS20659-165	25	8
	26	13
MS21044N06	7	24
MS21044N08 13 35		
MS21318-21	25	53
	26	66
MS21333-117	21	35
MS21333-126	21	23
MS21333-75	9	18
MS21333-76	9	17
	19	42
MS25036-101	18	16
MS25036-103	6	12
	18	12
	18	58
MS25036-104	18	18
MS25036-105	18	5
MS25036-106	18	68
MS25036-108	18	21
	18	57
MS25036-109	8	4A
	18	48
	18	13A
MS25036-110 18	6	
MS25036-112 18 10		
MS25036-113 18 20		
MS25036-116 18 6B		
MS25036-117 18 14		
MS25036-127 19		4
	19	8
	19	12
	19	15
MS25033-128 19		16
	19	20
	19	24
	19	28
MS25036-148	8	9
MS25036-149	8	8
	8	

PART NUMBER	FIG.	ITEM
MS25036-150	8	3
	18	9
MS25036-151	18	49
MS25036-152	8	157
MS25036-153	18	17
MS25036-154	8	87
	8	102
	8	133
	26	32
MS25036-157	26	49
MS25036-158	18	15
MS25041-5	6	7
MS25043-18DA	7	54
	19	63
MS25224-1	6	23
MS25251-12	7 86	
MS25251-16	8 30	
MS25281-4 25		27
	26	39
MS27183-13	25	45
	26	57
MS3102R18-11P	8	32
	18	91
MS3106F18-11S	7 83	
MS3367-1-9	7	82
	8	13
	13	48
	15	24
	15	59
	15	68
	18	26
	21	12
	21	43
	21	78
MS3367-4-9 15 67		
MS3367-5-9	7	21
	8	12
	18	25
MS3368-1-9A	8	14
	18	27
MS3420-10	7 85	
MS35190-253	3 25	
MS35198-42	9	2

PART NUMBER	FIG.	ITEM
MS35198-46	9 33	
MS35206-241	25	28
MS35206-248	26	36
MS35206-283	26	42
MS35206-287	26	4
MS35206-329	7 23	
MS35207-259	25	29
MS35333-38	26	38
MS35333-39	25	24
MS35333-40	26	30
MS35338-62	2	6
	3	2
	4	14
	5	4
	7	90
	15	56
MS35338-63	2	19
	3	29
	4	8
	10	5
	12	8
	13	26
	15	19
	20	33
	21	25
	26	3
	26	27
	27	6
MS35338-64	15	51
	25	15
MS35338-65	1	25
	13	30
	24	2
	29	2
MS35338-67	26	19
	27	18
MS35489-20	13	6
MS35489-27	19	48
MS35643-1 15	2	
MS35645-1 15	1	
MS35649-282	26	37
MS35650-103	21	60
MS35842-11	22	2

PART NUMBER	FIG.	ITEM
MS51007-12	7 79	
MS51007-6	7	80
MS51412-11	24	6
MS51412-2	2	7
	3	3
	4	15
	5	5
	7	29
	15	57
	19	32
MS51412-20	6 38	
MS51412-25	15	53
MS51412-4	2	20
	3	32
	10	7
	12	10
	13	28
	15	21
	19	50
	19	69
	20	34
	21	27
	27	8
MS51412-5	3	31
MS51412-7	1	19
	10	11
	13	45
	21	59
	27	2
	29	3
MS51412-8 15	4	
MS51412-9 27 14		
MS51481-06	13	34
MS51492-14	15	55
MS51493-2	2	48
	19	59
MS51493-3	2	49
MS51493-5 27	1	
MS51500A5-4 15 46		
MS51500A6 15 41		
MS51500B4-4Z	15	95
MS51504A6 15 79		
MS51512-2	7	91

PART NUMBER	FIG.	ITEM
MS51520A5 15 29		
MS51860-54Z	15	28
MS51922-2 19 68		
MS51922-9 25		41
	26	53
MS51937-5 25 50		
MS51937-7 26 63		
MS51943-13	24	5
	25	3
MS51943-9 27 13		
MS51967-14	25	51
MS51967-2 21		24
	26	29
	26	2
MS51967-20	26	64
MS51971-7 24 10		
MS52130A101360	BULK	13
MS52412-5	4	9
MS87006-11	13	10
MS87006-13	9	5
MS90725-110	26	18
MS90725-15	3 41	
MS90725-3 25		2
	26	6
	26	31
MS90725-34	25	14
MS90725-43	25	42
	26	54
MS90725-5	26	26
MS90725-6	25	9
	26	14
MS90725-58	25	44
	26	56
MS91523-2K2B	6	10
P00-2904	12	14
P10-1870	12	16
P12-028	12	19
P12-0307	12	18
P12-0316	12	15

PART NUMBER	FIG.	ITEM
RBB25	7	95
RR-C-271 TY2CL6 BULK	6	
RR-C-271 TY2 CL7-16	BULK	8
SAE-J1127 BULK	36	
SAE J1508 15 75		
SAE J1508-06	13	4
	15	11
SMP18-1072	12	17
Z60960	BULK	10
ZX-4134	BULK	28

END OF WORK PACKAGE

By Order of the Secretary of the Army:

GEORGE W. CASEY, JR
General, United States Army
Chief of Staff

Official:

JOYCE E. MORROW
Administrative Assistant to the
Secretary of the Army
1019414

By Order of the Secretary of the Air Force:

NORTON A SCHWARTZ
G *eneral, United States Air Force*
Chief of Staff

Official:

DONALD J HOFFMAN *General,*
United States Air Force
Commander, AFMC

DISTRIBUTION:

To be distributed in accordance with the initial distribution number (IDN) 255561 requirements for TM 9-6115-643-24P.

THE METRIC SYSTEM AND EQUIVALENTS

LINEAR MEASURE

1 Centimeter = 10 Millimeter = 0.01 Meters = 0.3937 inches

1 Meter = 100 Centimeters = 1000 Millimeters = 39.37 inches

1 kilometer = 1000 Meters = 0.621 Miles

SQUARE MEASURE

1 Sq. Centimeter = 100 Sq. Millimeter = 0.155 Sq. Inches

1 Sq. Meter = 10,000 Sq. Centimeters = 10.76 Sq. Inches

1 Sq. Kilometer = 1,000,000 Sq. Meters = 0.386 Sq. Miles

WEIGHTS

1 Gram = 0.0[illegible] Kilograms = 1000 Milligrams = 0.035 Ounces

1 Kilogram = 100 Grams = 2.2 lb.1 Cu. Meter = 1,000,000

1 Metric Ton = 1000 Kilograms = 1 Megagram = 1.1 Short Tons

CUBIC MEASURE

1 Cu. Centimeter = 1000 Cu. Millimeters = 0.06 Cu. Inches

1 Cu. Centimeters = 35.31 Cu. Feet

LIQUID MEASURE

1 Millimeter = 0.001 Liters = 0.0338 Fluid Ounces

1 Liter = 1000 Millimeters = 32.82 Fluid Ounces

TEMPERATURE

5/9 (°F - 32) = °C

212° Fahrenheit is equivalent to 100° Celsius

90° Fahrenheit is equivalent to 32.2° Celsius

32° Fahrenheit is equivalent to 0° Celsius

9/5 °C + 32 = °F

APPROXIMATE CONVERSION FACTORS

TO CHANGE	TO	MULTIPLY BY
Inches	Centimeters	2.540
Feet	Meters	0.305
Yards	Meters	0.914
Miles	Kilometers	1.609
Square Inches	Square Centimeters	6.451
Square Feet	Square Meters	0.093
Square Yards	Square Meters	0.836
Square Miles	Square Kilometers	2.590
Acres	Square Hectometers	0.405
Cubic Feet	Cubic Meters	0.028
Cubic Yards	Cubic Meters	0.765
Fluid Ounces	Milliliters	29.573
Pints	Liters	0.473
Quarts	Liters	0.946
Gallons	Liters	3.785
Ounces	Grams	28.349
Pounds	Kilograms	0.454
Short Tons	Metric Tons	0.907
Pound-Feet	Newton-Meters	1.356
Pounds per Square Inch	Kilo pascals	6.895
Miles per Gallon	Kilometers per Liter	0.425
Miles per Hour	Kilometers per Hour	1.609

TO CHANGE	TO	DIVIDE BY
Centimeters	Inches	2.540
Meters	Feet	0.305
Meters	Yards	0.914
Kilometers	Miles	1.609
Square Centimeters	Square Inches	6.451
Square Meters	Square Feet	0.093
Square Meters	Square Yards	0.836
Square Kilometers	Square Miles	2.590
Square Hectometers	Acres	0.405
Cubic Meters	Cubic Feet	0.028
Cubic Meters	Cubic Yards	0.765
Milliliters	Fluid Ounces	29.573
Liters	Pints	0.473
Liters	Quarts	0.946
Liters-Meters	Gallons	3.785
Grams	Ounces	28.349
Kilograms	Pounds	0.454
Metric Tons	Short Tons	0.907
Newton-Meters	Pound-Feet	1.356
Kilo pascals	Pounds per Square Inch	6.895
Kilometers per Liter	Miles per Gallon	0.425
Kilometers per Hour	Miles per Hour	1.609

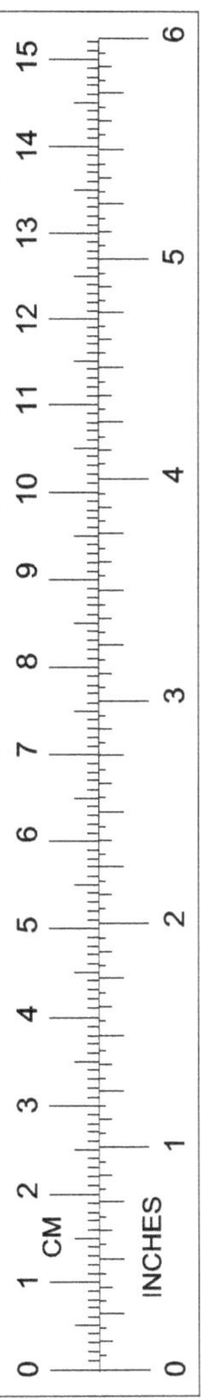

PIN:071610-000

www.ingramcontent.com/pod-product-compliance
Lightning Source LLC
Chambersburg PA
CBHW080417030426
42335CB00020B/2477